WORKBOOK
MODERN
METALWORKING

by

John R. Walker

Publisher

The Goodheart-Willcox Company, Inc.
Tinley Park, Illinois

HOW TO USE THIS WORKBOOK

This workbook is designed for use with the textbook *Modern Metalworking*. It has been planned to help you develop a better understanding of the metalworking industries, related occupations, and career opportunities. Whether or not you intend to work in one of these industries, you will still be dependent upon them in some way. Just about every manufactured product uses metal somewhere in its processing, production, or construction.

To get the most out of this workbook, it will be necessary for you to carefully study each chapter in the textbook before attempting the related workbook assignments. Pay particular attention to the illustrations. They have been specially prepared or selected to help you more fully comprehend the *Modern Metalworking* textbook. Also take personal note of the material on safety. Many shop accidents are caused because of carelessness or ignorance of safety regulations.

The *Modern Metalworking Workbook* has six types of questions. They are: completion, listing, multiple choice, matching, identification, and short answer. In addition, some chapters have drawing or sketching assignments.

After reading the textbook assignment, carefully study the related questions in the workbook. Write the best answer in the provided space. Make sketches where required, construct projects, and complete activities as directed. Try to complete as much of each assignment as you can without reviewing the textbook.

The world of technology is becoming very complex. There is very little need for unskilled workers. The more you know about metalworking, the better your chances will be to secure employment in this field. You will be doing work you enjoy and, at the same time, helping to make the world a better place in which to live.

John R. Walker

TABLE OF CONTENTS

TECHNOLOGY AND CAREERS

1

Name: _____   **Date:** _____

Instructor: _____   **Score:** _____

Carefully study the chapter. Then answer the following questions.

PART I

1. _____ has done more to shape the world we live in than all other forces.

   1. _____

2. _____ was made possible by the first use of tools.

   2. _____

3. Slightly more than two hundred years ago, changes started to occur that completely altered the way people lived. This eventually became known as the _____.

   3. _____

4. List several changes that occurred during the time mentioned in question 3. _____

   _____

   _____

   _____

   _____

5. We are now in what is known as a technological revolution. List some recent technological accomplishments.

   _____

   _____

   _____

   _____

6. We know the careless use of technology has been responsible for some environmental problems. How will these problems be corrected or solved?

   6. _____

   (a) By the government.
   (b) By innovative technology.
   (c) By themselves.
   (d) All of the above.
   (e) None of the above.

7. Name four types of technology. _____

_____

_____

8. Name two areas of technology *not* described in the text. _____

_____

_____

9. _____ is the income of a business after all expenses have   9. _____
   been paid.

10. Why is profit necessary? _____

_____

_____

_____

_____

11. The person in the metalworking industry who performs   11. _____
    operations that do not require a high degree of skill or
    training is classified as a(n) _____ worker.

12. Trained individuals who are capable of performing the   12. _____
    exacting work and skills of a trade are called _____ workers.

13. What is an *apprentice program?* _____

_____

_____

_____

_____

14. One of the jobs of the _____ is to assist the engineer   14. _____
    by constructing and testing experimental devices and
    equipment.

15. The mechanical engineer is usually responsible for the   15. _____
    _____ and _____ of new machines, devices, and ideas.

16. An industrial engineer is primarily concerned with the   16. _____
    safest and most efficient use of _____.
    (a) machines
    (b) materials
    (c) personnel
    (d) All of the above.
    (e) None of the above.

Name_____

17. The tool and manufacturing engineer is primarily concerned with the _____.
    (a) design and development of original or prototype models
    (b) methods and means needed to manufacture and assemble a product
    (c) safety of the finished product
    (d) None of the above.
    (e) All of the above.

17._____

18. The _____ engineer designs and develops aircraft and aerospace vehicles.

18._____

19. _____ engineers are responsible for the development and testing of metals used in products or manufacturing processes.

19._____

20. _____ consists of people who plan, direct, and supervise the operation of an industrial organization.

20._____

21. Why does American technology require strong and dynamic leadership?_____

_____

_____

_____

22. The quality of leadership usually determines whether an organization will be a(n) _____ or a(n) _____.

22._____

PART II

23. What does industry expect of you as an employee? _____

_____

_____

_____

_____

_____

24. Where can information on metalworking occupations be found? _____

_____

_____

_____

_____

25. Describe how you would prepare for, and go about getting, a job. _____

_____

_____

_____

_____

_____

_____

_____

26. We know that just about every "worker-made" product has used metal somewhere in its manufacture. List as many occupations and businesses as you can think of that use metal directly in one
of its many forms, or indirectly as tools and equipment.

_____

_____

_____

_____

_____

_____

_____

27. List three factors that can lead to rejection to employment (not getting the job). _____

_____

_____

_____

28. What are three problems that can lead to job termination (getting fired)? _____

_____

_____

_____

29. What occupations or businesses can you think of that do *not* use metal in some shape or form either
directly or indirectly in their operation?

_____

_____

_____

_____

_____

Name_____

PART III

30. The following job application has been included to help you gather the general information required on most employment applications. Fill it out as completely as you can, making sure the information is correct. Write N/A if it does not apply to you.

APPLICATION FOR EMPLOYMENT

PERSONAL INFORMATION

Date_____   Social Security Number_____

Name_____
   Last   First   Middle

Present Address_____
         Street   City   State

Permanent Address_____
         Street   City   State

Phone No._____

If related to anyone in our employ,   Referred
state name and department_____   by_____

EMPLOYMENT DESIRED

Position_____   Date you can start_____   Salary desired_____

Are you employed now?_____   If so may we inquire of your present employer?_____

Ever applied to this company before?_____   Where_____   When_____

EDUCATION

	Name and Location of School	Years Completed	Subjects Studied
Grammar School			
High School			
Trade, Business or Correspondence School			

Subject of special study or research work_____

What foreign languages do you speak fluently? _____ Read fluently? _____ Write fluently? _____

U.S. Military
service _____   Rank _____   Present membership in
National Guard or Reserves _____

_____

FORMER EMPLOYERS List below last three employers starting with last one first

Date Month and Year	Name and Address of Employer	Salary	Position	Reason for Leaving
From ____				
To				
From ____				
To				
From ____				
To				

REFERENCES Give below the names of two persons not related to you whom you have known for at least one year

	Name	Address	Job Title	Years Acquainted
1				
2				

Name_____

PART IV

31. Look through the help wanted sections of your local newspapers. Cut out ads relating to the metalworking industry and paste them to this page. You may also search job sites on the Internet. Print this information and paste it to this page.

JOBS RELATING TO THE METALWORKING INDUSTRY

CLASSIFYING METALS

2

Name: _____   Date: _____

Instructor: _____   Score: _____

Carefully study the chapter. Then answer the following questions.

PART I

1. Some metals have unusual characteristics. How many metals can you list? What are their unusual characteristics?

   _____   _____

   _____   _____

   _____   _____

   _____   _____

   _____   _____

2. How do ferrous and nonferrous metals differ? _____

   _____

   _____

   _____

3. What is a *base metal?* _____

   _____

4. What is an *alloy?* _____

   _____

5. List three precautions that should be observed when handling metals.

   1) _____

   _____

   2) _____

   _____

   3) _____

   _____

6. Industry uses more than a thousand different metals. Using the partial list below, separate the metals into ferrous or nonferrous categories.

cast iron	manganese	tungsten carbide	steel
tungsten	magnesium	wrought iron	pewter
beryllium	aluminum	lead	columbium
copper	brass	low-carbon steel	tin
zinc	tantalum	stainless steel	bronze
gold	chromium	nickel	silver
carbon steel	molybdenum	high-carbon steel	

Ferrous Metals: _____

_____

_____

_____

_____

_____

Nonferrous Metals: _____

_____

_____

_____

_____

_____

7. Further classify the metals listed in question 6 into base metal or alloy categories.

Base Metals: _____

_____

_____

_____

_____

_____

Alloys: _____

_____

_____

_____

_____

_____

8. Briefly describe the following metals and their uses.

Wrought iron: _____

_____

Name_____

Carbon steel: _____

_____

Tungsten carbide: _____

_____

Magnesium: _____

_____

Brass: _____

_____

Tantalum: _____

_____

9. What is unusual about mercury? _____

_____

_____

10. What is a *honeycomb structure?* What purpose does it serve?_____

_____

_____

_____

_____

11. What are *composites?* _____

_____

_____

_____

_____

12. _____ steels are classified by their percentage of carbon in points.   12._____

Identify the following categories of metals.

13. Metals containing iron._____

14. Metals containing no iron. _____

15. A mixture of two or more metals. _____

16. Metals like copper, tin, and zinc. _____

17. Metals like gold, silver, and platinum. _____

PART II

Match each definition with the correct term.

_____18. Protective coating on galvanized steel and iron sheet.

_____19. A substitute for silver in jewelry making.

_____20. An alloy of copper and tin.

_____21. An alloy of copper and zinc.

_____22. A soft, shiny, silvery metal.

_____23. An alloy of iron and other metallic elements.

_____24. Contains chromium and is corrosion resistant.

_____25. Purifies steel and adds strength and toughness.

_____26. Hardest human made metal.

_____27. The lightest of structural metals.

_____28. As strong as steel, but only half as heavy.

_____29. Has a black oxide coating.

_____30. Technique employed to determine grades of steel.

_____31. Oldest metal known.

_____32. Used in jewelry and automobile catalytic converters.

_____33. Also known as Britannia metal.

_____34. Tools made of this metal can make deeper cuts at higher machine speeds than regular tool steels.

(a) Steel
(b) Manganese
(c) High-speed steel
(d) Tungsten carbide
(e) Stainless steel
(f) Hot-rolled steel
(g) Spark test
(h) Magnesium
(i) Titanium
(j) Brass
(k) Bronze
(l) Copper
(m) Tin
(n) Zinc
(o) Pewter
(p) German silver
(q) Platinum

PART III

35. Examine the metal samples your instructor will pass among you. Identify them by placing the name of the metal in the appropriately numbered space.

1. _____

2. _____

3. _____

4. _____

5. _____

6. _____

7. _____

8. _____

9. _____

10. _____

UNDERSTANDING DRAWINGS

3

Name: _____   Date: _____

Instructor: _____   Score: _____

Carefully study the chapter. Then answer the following questions.

PART I

1. Why are drawings used in the metalworking industry? _____

   _____

   _____

   _____

2. What are dimensions and why are they provided on a drawing? _____

   _____

   _____

   _____

   _____

3. What type of drawing includes both US Conventional and metric dimensions? _____

   _____

4. What are *tolerances*? What do they indicate on a drawing? _____

   _____

   _____

   _____

   _____

   _____

5. When a tolerance is shown on the drawing as plus *and* minus ($2\ 1/2^{\pm 1/64}$), it is called a(n) _____ tolerance.

   5. _____

6. When a tolerance is shown on the drawing as plus *or* minus ($2\ 1/2^{+1/64}$ or $2\ 1/2^{-1/64}$), it is called a(n) _____ tolerance.

   6. _____

7. Why are drawings numbered? _____

_____

_____

8. What type of drawing shows the part in one or more views and includes all of the dimensions and information needed to make it?

8. _____

9. What type of drawing shows where and how the part shown in the drawing fits into the complete assembly of the unit?

9. _____

10. Why was *Geometric Dimensioning and Tolerancing* developed? _____

_____

_____

_____

_____

11. Why should your work be planned? _____

_____

_____

_____

_____

12. What is a *plan of procedure?* _____

_____

_____

_____

_____

13. What should be the first step in planning a project? _____

_____

_____

_____

_____

14. What does a *bill of materials* include? _____

_____

_____

_____

_____

PART II

15. Lines, symbols, and special figures are used to give meaning to a drawing. The drawing shown below is typical of those used by industry. In the appropriate blank, write the name of the line and how it is used.

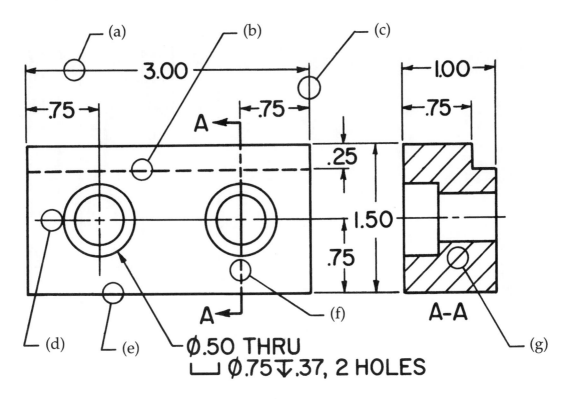

Line	How it is used
(a) _____	_____
(b) _____	_____
(c) _____	_____
(d) _____	_____
(e) _____	_____
(f) _____	_____
(g) _____	_____

PART III

Carefully study the working drawing shown. Then answer the following questions.

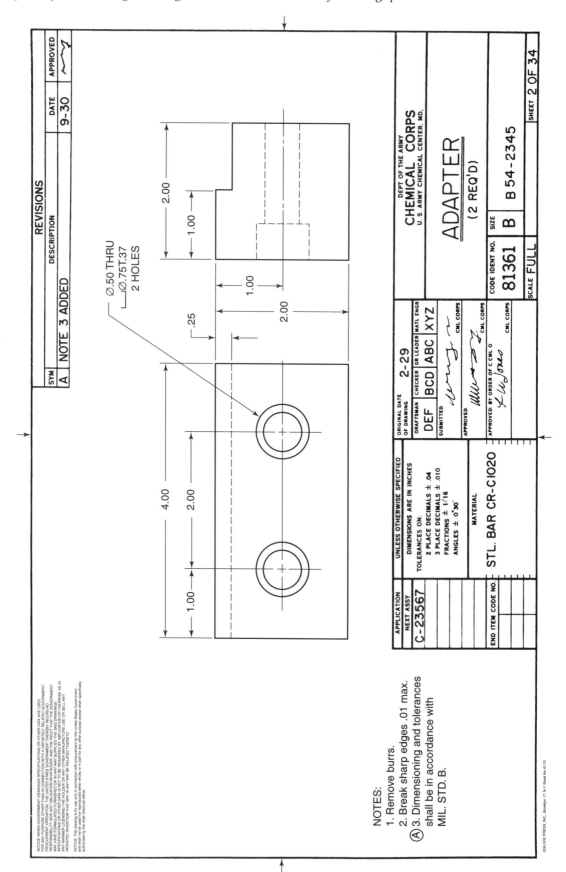

Name_____

16. What is the name of the part?

16._____

17. What is the number assigned to the drawing?

17._____

18. What is the scale of the original drawing?

18._____

19. The part is made of what material?

19._____

20. What tolerances are allowed for fractional dimensions?

20._____

21. What is the length of the part?

21._____

22. The part is _____ high and _____ thick.

22._____

_____

23. How many holes are drilled in the block?

23._____

24. What is the diameter of the holes?

24._____

25. The centerline for the holes is _____ from the bottom edge of the part.

25._____

26. The centerline of one hole is _____ from the left edge of the part.

26._____

27. The centerline for the second hole is _____ from the left edge of the part.

27._____

28. Both holes are counterbored _____ deep.

28._____

29. The size of the shoulder cut in the top surface of the part is _____ long, _____ wide, and _____ deep.

29._____

_____

_____

30. What special information concerning the finish on the part is given on the drawing?

_____

_____

_____

_____

_____

SAFETY PRACTICES

Name: _____   Date: _____

Instructor: _____   Score: _____

Carefully study the chapter. Then answer the following questions.

PART I

1. Name three types of eye protection. _____
   _____
   _____

2. What is the cause of most accidents? _____
   _____

3. Why should all types of jewelry be removed before performing work in the metalworking lab/shop?
   _____
   _____

4. For what types of metalworking applications should a mask be worn? _____
   _____
   _____
   _____

5. Why is it important to keep the metalworking area clean? _____
   _____
   _____
   _____

6. Portable power tools should *not* be operated in areas where thinners and solvents are in use because a serious _____ or _____ could result.

   6. _____
   _____

7. There are many safety rules that should be kept in mind while doing any kind of shop work. List five of them.

1) _____
   _____

2) _____
   _____

3) _____
   _____

4) _____
   _____

5) _____
   _____

8. List five safety rules that must be followed when operating machinery.

1) _____
   _____

2) _____
   _____

3) _____
   _____

4) _____
   _____

5) _____
   _____

9. All of us have seen unsafe actions in the shop. List several that you have observed. Do *not* use the names of the students who are responsible for the unsafe actions.

_____

_____

_____

_____

PART II

10. Space is provided on the next two pages for you to design safety posters. They should be in color. Select two topics from the following list:

   (a) Always wear your safety glasses.
   (b) No horseplay in the shop/lab.
   (c) The ABC's of safety—Always Be Careful.
   (d) It hurts to get hurt.
   (e) A safety theme of your own choosing.

Name_____

MEASUREMENT

Name: _____   **Date:** _____

Instructor: _____   **Score:** _____

Carefully study the chapter. Then answer the following questions.

PART I

1. The _____ part of a standard inch is known as a microinch.

   1. _____

2. Steel rules are useful for measuring to _____ inches and _____ mm.

   2. _____

   _____

3. Identify the following rules.

   (a) _____

   (b) _____

   (c) _____

4. Make readings from the rule shown below. Place the answers in the proper blank.

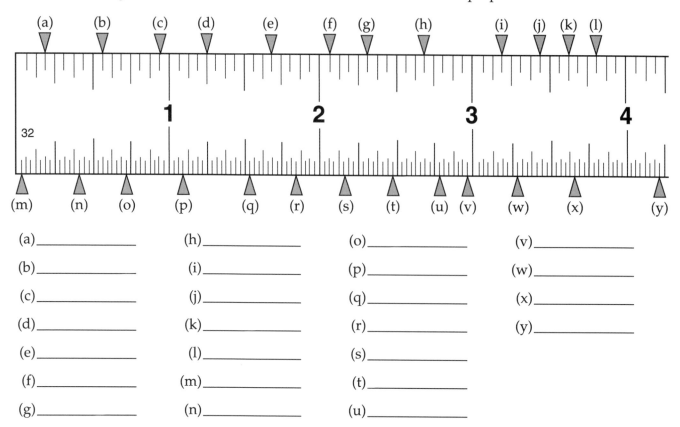

(a)_____   (h)_____   (o)_____   (v)_____

(b)_____   (i)_____   (p)_____   (w)_____

(c)_____   (j)_____   (q)_____   (x)_____

(d)_____   (k)_____   (r)_____   (y)_____

(e)_____   (l)_____   (s)_____

(f)_____   (m)_____   (t)_____

(g)_____   (n)_____   (u)_____

5. Make readings from the rule shown below. Place your answers in the proper blank.

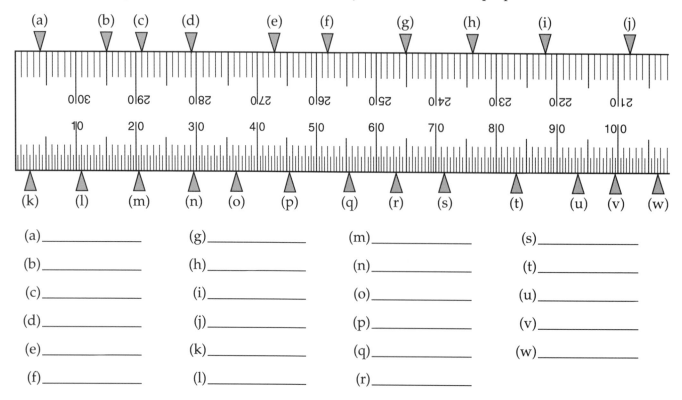

(a)_____   (g)_____   (m)_____   (s)_____

(b)_____   (h)_____   (n)_____   (t)_____

(c)_____   (i)_____   (o)_____   (u)_____

(d)_____   (j)_____   (p)_____   (v)_____

(e)_____   (k)_____   (q)_____   (w)_____

(f)_____   (l)_____   (r)_____

PART II

6. Make readings from the micrometers shown below and place your answers in the proper blank.

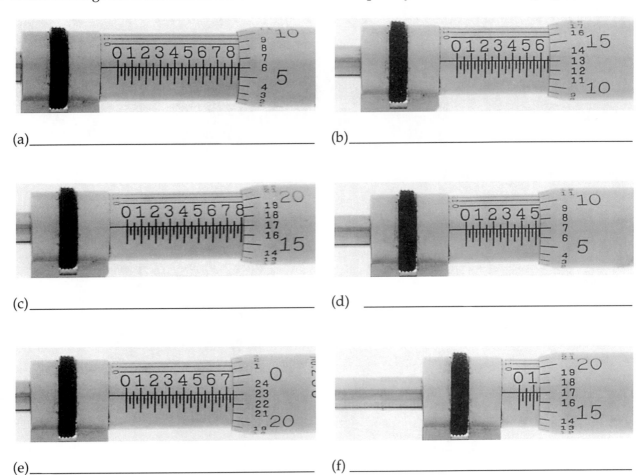

(a)_____

(b)_____

(c)_____

(d) _____

(e)_____

(f) _____

7. Make readings from the micrometers shown below and place your answers in the proper blank.

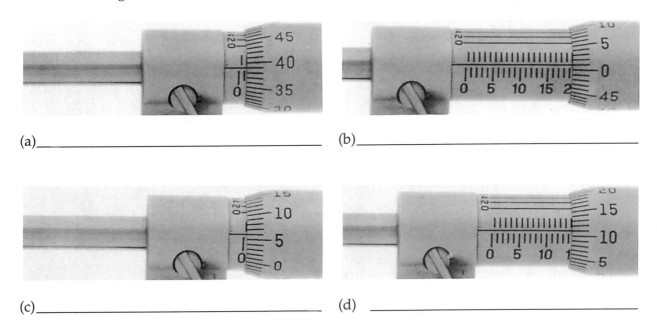

(a)_____

(b)_____

(c)_____

(d) _____

8. Identify the following measuring tools. Write the name of each tool in the correct space.

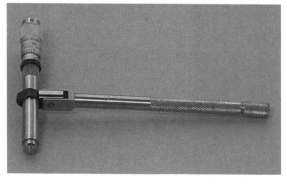

(a) _____

(b) _____

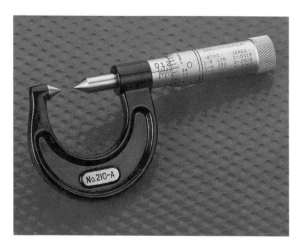

(c) _____

9. A micrometer depth gage can be used to measure the depth of holes, slots, and projections. How does reading the measurements on this tool differ from reading measurements on a micrometer caliper?

_____

_____

_____

_____

10. _____ aid in determining the pitch diameter of screw threads.

10. _____

11. The _____ has an advantage over the micrometer in that it can be used to make both inside and outside measurements over a range of sizes.

11. _____

12. _____ calipers provide dial readings that are combined with readings from a beam.

12. _____

Name_____

PART III

13. Make readings from the Vernier calipers shown below. Place your answers in the proper blanks.

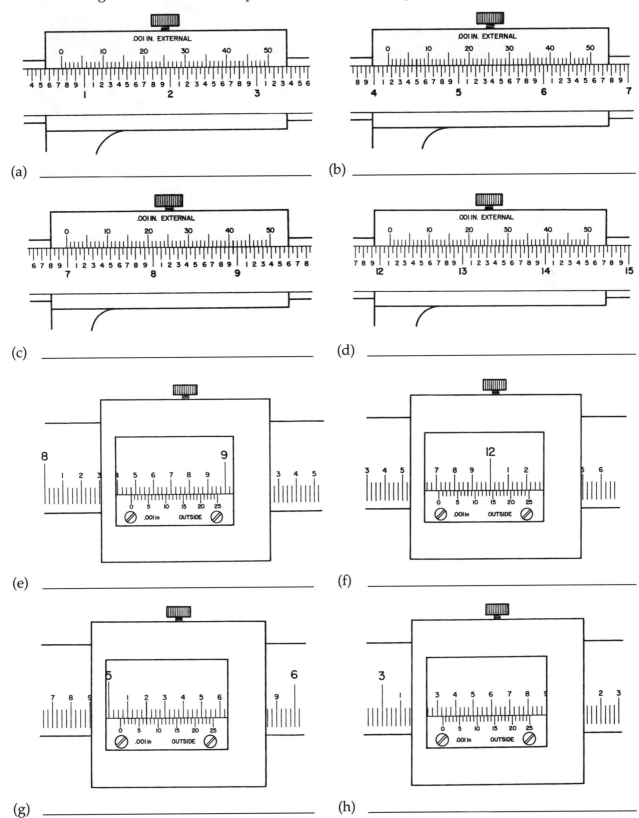

(a) _____

(b) _____

(c) _____

(d) _____

(e) _____

(f) _____

(g) _____

(h) _____

14. Identify the following measuring tools. Write the name of each tool in the correct space.

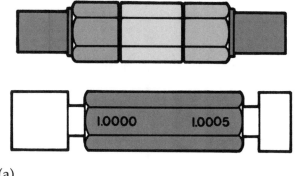

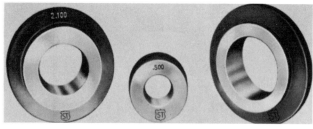

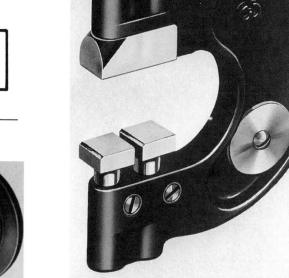

(a) _____

(b) _____   (c) _____

15. The plug gage is used to check hole diameters that _____.   15. _____

   (a) are within specified tolerances
   (b) cannot be measured by any other method
   (c) are too small for other measuring tools
   (d) None of the above.
   (e) All of the above.

PART IV

16. The ring gage is used to check _____ (external/internal)   16. _____
    shaft diameters.

17. Name some uses of the dial indicator. _____

    _____

    _____

    _____

18. Thickness gages are ideal for _____.   18. _____

   (a) measuring narrow slots
   (b) setting small gaps
   (c) determining fit between mating surfaces
   (d) All of the above.
   (e) None of the above.

Name_____

19. Identify the following measuring tools. Write the name of each tool in the correct space.

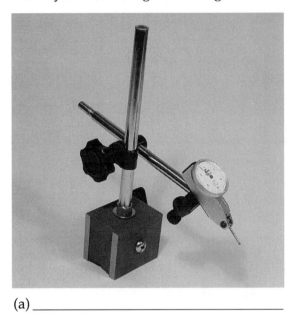

(b) _____

(a) _____

20. Identify the following measuring tools. Write the name of each tool in the correct space.

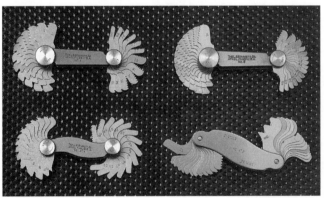

(a) _____

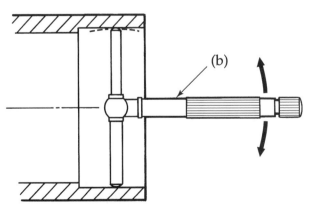

(b) _____

(c) _____

21. Convex and concave radii on corners or against shoulders can be checked with a(n) _____ gage.

21. _____

22. A(n) _____ is a tool that can be used to make accurate measurements but must be used with a micrometer or Vernier caliper.

22. _____

    (a) plug gage
    (b) telescoping gage
    (c) dial indicator
    (d) ring gage

PART V

The following exercises will aid you in your understanding of the metric system and give you practice in converting from one system to the other. Use the tables on pages 37 and 38 to work the problems below. Study the examples, then complete the following exercises. Show your work. List your answers to a maximum of two decimal places.

    Example: 1/2″ = 12.7 mm from the table on page 38.

    Example: 3″ = 76.2 mm (3 × 25.4 = 76.2 mm) from the table on page 37.

23. 1/8″ = _____ cm

24. 15/16″ = _____ mm

25. 3/8″ = _____ cm

26. 15/32″ = _____ mm

27. 2″ = _____ mm

28. 2 cm = _____ inches

29. 11/16″ = _____ m

30. 5/8″ = _____ cm

31. 51/64″ = _____ mm

32. 6.5 mm = _____ cm

33. 1.27 cm = _____ mm

34. 0.125″ = _____ mm

35. 0.875″ = _____ cm

36. 3.8125″ = _____ mm

37. 6 gallons = _____ Liters

38. 3 lb. = _____ kg

39. 44 cu. inches = _____ I

40. 10 miles = _____ km

METRIC SYSTEM

The basic unit of the metric system is the meter (m). The meter is exactly 39.37″ long. This is 3.37″ longer than the US Customary yard. Units that are multiples or fractional parts of the meter are designated as such by prefixes to the word meter. For example:

1 millimeter (mm) = 0.001 meter or 1/1000 meter
1 centimeter (cm) = 0.01 meter or 1/100 meter
1 decimeter (dm) = 0.1 meter or 1/10 meter
1 meter (m)
1 decameter (dkm) = 10 meters
1 hectometer (hm) = 100 meters
1 kilometer (km) = 1000 meters

These prefixes may be applied to any unit of length, weight, volume, etc. The meter is adopted as the basic unit of length, the gram for mass, and the liter for volume.

Name_____

In the metric system, area is measured in square kilometers (km^2), square centimeters (cm^2), etc. Volume is commonly measured in cubic centimeters, etc. One liter (L) is equal to 1000 cubic centimeters.

The metric measurements in most common use are shown in the following tables:

Measures of Length
10 millimeters = 1 centimeter
10 centimeters = 1 decimeter
10 decimeters = 1 meter
1000 meters = 1 kilometer (km)

Measures of Weight
1000 milligrams = 1 gram (g)
1000 grams = 1 kilogram (kg)
1000 kilograms = 1 metric ton

Measures of Volume
1000 cubic centimeters = 1 liter (L)
100 liters = 1 hectoliter

CONVERSION TABLE

To convert   **Multiply by**

Length

km to miles	0.62
miles to km	1.61
meters to miles	0.00062
miles to meters	1609.35
meters to yards	1.0936
yards to meters	0.9144
cm to inches	0.3937
inches to cm	2.54
mm to inches	0.03937
inches to mm	25.4

Volume

cm^3 to cubic inches	0.061
cubic inches to cm^3 or ml	16.387
liters to cubic inches	61.024
cubic inches to liters	0.0164
liters to gallons	0.264
gallons to liters	3.785

Weight

kg to lb.	2.2
lb. to kg	0.4536
grams to oz.	0.0353
oz. to grams	28.35

METRIC — INCH EQUIVALENTS

INCHES Fractions	INCHES Decimals	MILLIMETERS	INCHES Fractions	INCHES Decimals	MILLIMETERS
	.00394	.1	15/32	.46875	11.9063
	.00787	.2		.47244	12.00
	.01181	.3	31/64	.484375	12.3031
1/64	.015625	.3969	1/2	.5000	12.70
	.01575	.4		.51181	13.00
	.01969	.5	33/64	.515625	13.0969
	.02362	.6	17/32	.53125	13.4938
	.02756	.7	35/64	.546875	13.8907
1/32	.03125	.7938		.55118	14.00
	.0315	.8	9/16	.5625	14.2875
	.03543	.9	37/64	.578125	14.6844
	.03937	1.00		.59055	15.00
3/64	.046875	1.1906	19/32	.59375	15.0813
1/16	.0625	1.5875	39/64	.609375	15.4782
5/64	.078125	1.9844	5/8	.625	15.875
	.07874	2.00		.62992	16.00
3/32	.09375	2.3813	41/64	.640625	16.2719
7/64	.109375	2.7781	21/32	.65625	16.6688
	.11811	3.00		.66929	17.00
1/8	.125	3.175	43/64	.671875	17.0657
9/64	.140625	3.5719	11/16	.6875	17.4625
5/32	.15625	3.9688	45/64	.703125	17.8594
	.15748	4.00		.70866	18.00
11/64	.171875	4.3656	23/32	.71875	18.2563
3/16	.1875	4.7625	47/64	.734375	18.6532
	.19685	5.00		.74803	19.00
13/64	.203125	5.1594	3/4	.7500	19.05
7/32	.21875	5.5563	49/64	.765625	19.4469
15/64	.234375	5.9531	25/32	.78125	19.8438
	.23622	6.00		.7874	20.00
1/4	.2500	6.35	51/64	.796875	20.2407
17/64	.265625	6.7469	13/16	.8125	20.6375
	.27559	7.00		.82677	21.00
9/32	.28125	7.1438	53/64	.828125	21.0344
19/64	.296875	7.5406	27/32	.84375	21.4313
5/16	.3125	7.9375	55/64	.859375	21.8282
	.31496	8.00		.86614	22.00
21/64	.328125	8.3344	7/8	.875	22.225
11/32	.34375	8.7313	57/64	.890625	22.6219
	.35433	9.00		.90551	23.00
23/64	.359375	9.1281	29/32	.90625	23.0188
3/8	.375	9.525	59/64	.921875	23.4157
25/64	.390625	9.9219	15/16	.9375	23.8125
	.3937	10.00		.94488	24.00
13/32	.40625	10.3188	61/64	.953125	24.2094
27/64	.421875	10.7156	31/32	.96875	24.6063
	.43307	11.00		.98425	25.00
7/16	.4375	11.1125	63/64	.984375	25.0032
29/64	.453125	11.5094	1	1.0000	25.4001

LAYOUT WORK

Name: _____ **Date:** _____

Instructor: _____ **Score:** _____

Carefully study the chapter and then answer the questions.

PART I

1. _____ is a term used to describe the locating and marking out of lines, circles, arcs, and centerpoints for drilling holes.

   1._____

2. Layout lines are easier to see if _____ is first applied.

   2._____

3. What tool is illustrated below? What purpose does it serve? _____

   _____

   _____

4. Circles and arcs are drawn on metal surfaces with _____.

   4._____

5. Circles and arcs that are too large to be drawn with the above tool are drawn with the _____.

   5._____

6. What tool is illustrated below? What purpose does it serve? _____

   _____

   _____

7. Identify the tools illustrated below. Write their names in the spaces provided.

(a) _____

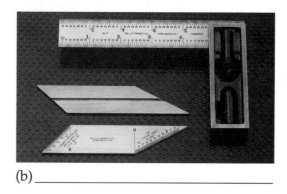

(b) _____

(c) _____

8. For what purpose(s) is a square used? _____

_____

_____

_____

9. Name at least three uses for a combination set. _____

_____

_____

_____

10. Identify the tools illustrated below. What purpose do they serve? _____

_____

_____

_____

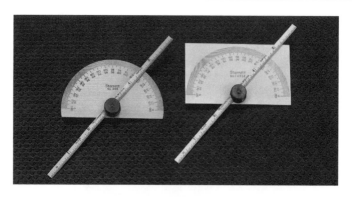

Name_____

PART II

11. The precise flat surface employed for layout work and inspection is called a(n) _____.

11._____

12. Round stock is frequently supported on _____ for layout and inspection purposes.

    (a) parallels
    (b) angle plates
    (c) V-blocks
    (d) None of the above.

12._____

13. A(n) _____ is often used to check the accuracy of flat surfaces and as a guide for drawing long, straight lines.

13._____

14. Sketch in the points of the punches drawn below.

Center Punch

Prick Punch

15. List four safety practices that should be observed when doing layout work.

    1) _____

    _____

    2) _____

    _____

    3) _____

    _____

    4) _____

    _____

HAND TOOLS

7

Name: _____ Date: _____

Instructor: _____ Score: _____

Carefully study the chapter. Then answer the following questions.

PART I

1. What tool is illustrated below? _____

Match the descriptions with the tools listed.

_____ 2. Handy when space is limited and for holding small work.

_____ 3. Clamping device used to hold and position material while it is worked.

_____ 4. Also known as slip-joint pliers.

_____ 5. Designed to cut flush with work surface.

_____ 6. Useful for cutting heavier wire and pins.

(a) Vise
(b) Combination pliers
(c) Side-cutting pliers
(d) Needle-nose pliers
(e) Diagonal pliers

7. Identify the pliers illustrated below.

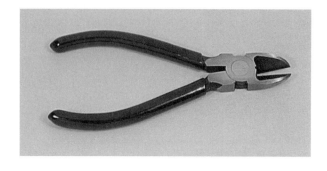

(a) _____

(b) _____

(c) _____

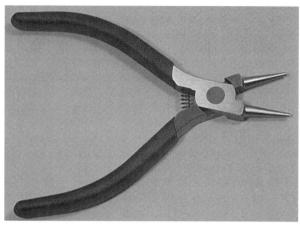

(d) _____

8. Identify the wrenches illustrated below.

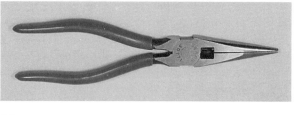

(a) _____

(b) _____

(c) _____

Name_____

PART II

Match the descriptions with the tools listed.

_____ 9. Has a hardened striking face.

_____ 10. Double-ended with two different size openings.

_____ 11. Box-like and made to fit many types of handles.

_____ 12. Family of tools used for assembling and disassembling threaded fasteners.

_____ 13. Permits tightening a threaded fastener for maximum holding power without danger of the fastener failing.

_____ 14. Can be adjusted to fit different size bolt heads and nuts.

_____ 15. Wrench used to grip round stock.

_____ 16. Wrench opening completely surrounds the bolt head or nut.

_____ 17. Will not rebound like other hammers.

_____ 18. Designed to turn flush and recessed type threaded fasteners.

_____ 19. Wrench with one open end and one box end.

_____ 20. Used with socket headed fasteners.

(a) Wrenches
(b) Torque-limiting wrench
(c) Adjustable wrench
(d) Pipe wrench
(e) Open-end wrench
(f) Box wrench
(g) Combination wrench
(h) Socket wrench
(i) Allen wrench
(j) Spanner wrench
(k) Ball-peen hammer
(l) Dead blow hammer
(m) Screwdrivers

PART III

21. List four general safety rules for wrench use.

   1) _____
   _____

   2) _____
   _____

   3) _____
   _____

   4) _____
   _____

22. Identify the screwdriver heads illustrated on the right.

(a) _____

(b) _____

(c) _____

(d) _____

(e) _____

(f) _____

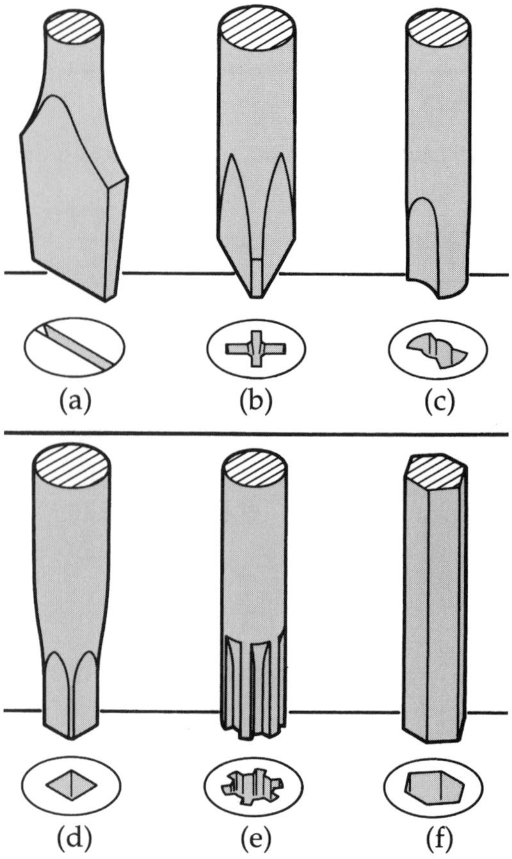

(a) (b) (c)

(d) (e) (f)

23. List three general safety precautions for screwdriver use.

1) _____

_____

2) _____

_____

3) _____

_____

24. List four general safety rules for striking tool use.

1) _____

_____

2) _____

_____

3) _____

_____

4) _____

_____

Name_____

PART IV

25. All of us have seen unsafe use of hand tools. List several that you have observed in the shop/lab. Do *not* use the names of students who were responsible for the unsafe actions.

_____

_____

_____

_____

_____

_____

HAND TOOLS
THAT CUT

Name: _____ **Date:** _____

Instructor: _____ **Score:** _____

Carefully study the chapter. Then answer the following questions.

PART I

1. Prepare sketches of the four basic types of chisels. Label each sketch.

_____

_____

_____

_____

2. For cutting on a flat plate, the chisel should have a(n) _____ cutting edge. When shearing metal held in a vise, the cutting edge should be _____.

   2. _____

   _____

3. A special chisel used for removing rivets is called a(n) _____.

   3. _____

4. List three safety precautions that should be observed when using chisels.

1) _____

_____

2) _____

_____

3) _____

_____

PART II

5. A hacksaw blade must be fitted in the frame so it will cut on the _____ stroke.

5. _____

6. Cutting work not mounted solidly or close to the vise will cause the work to _____, causing the blade to dull.

6. _____

7. What will be the result if a new blade is started in a cut already made by a dull blade? How can this problem be avoided?

_____

_____

_____

8. As a hacksaw blade becomes dull, the _____ becomes narrower.

8. _____

9. How can thin metal be cut with a hacksaw? _____

_____

_____

10. List four safety precautions relating to hacksaw safety.

1) _____

_____

2) _____

_____

3) _____

_____

4) _____

_____

Name_____

PART III

11. Identify the file cuts illustrated below.

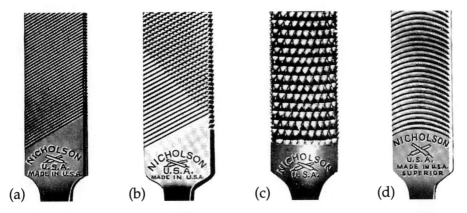

   (a)   (b)   (c)   (d)

(a) _____

(b) _____

(c) _____

(d) _____

12. Some soft metals cause _____; that is, the file teeth become loaded with some of the material that has been removed.

12. _____

13. Files should be cleaned frequently with a(n) _____ or a(n) _____.

13. _____

        _____

14. Files have three distinct characteristics. List them. _____

_____

_____

_____

15. Sketch the cross sections of the following files.

       3-square   Rat tail   Crossing

16. When a file is pushed lengthwise, straight ahead, or at a slight angle across the work, the operation is called _____ filing.

16._____

17. When a file is pushed and pulled across the work, the operation is called _____ filing.

17._____

18. List three safety precautions relating to filing.

1) _____

_____

2) _____

_____

3) _____

_____

PART IV

19. When is a hand reamer used? _____

_____

_____

_____

_____

20. When is an expansion reamer used? _____

_____

_____

_____

_____

21. The fastest way to ruin the cutting edges on a hand reamer is to turn it in a(n) _____ direction in the work.

21._____

22. List three safety precautions that should be observed when using hand reamers. _____

_____

_____

_____

_____

_____

_____

Name_____

PART V

23. Examine the chisels used in your shop. Select one for sharpening. Grind the head so there is no danger of chips flying from it and injuring someone.

24. Cut a piece of 1/8" × 1" section of metal in a vise, using the correct chisel.

25. Prepare two rectangular blocks of aluminum 2" × 1" × 1/2". Clamp them together. Divide the length of metal into three equal parts. At the first division on the joint, drill a 1/2" diameter hole. At the second division, drill the hole for a 1/2" diameter reamer and hand ream it. Be sure to use the proper cutting fluid for each operation. When separated, the two blocks illustrate the difference between a drilled and a reamed hole.

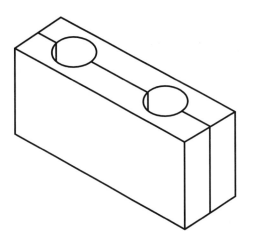

HAND THREADING

Name: _____ **Date:** _____

Instructor: _____ **Score:** _____

Carefully read the chapter. Then answer the following questions.

PART I

1. For a given diameter, how does the UNF series differ from the UNC series thread? _____
   _____
   _____

2. A tap is used to cut _____ threads. 2._____

3. What taps must be used to cut threads to the bottom of a blind hole? _____
   _____

4. A _____ tap is used to start threads. 4._____
   (a) taper
   (b) bottoming
   (c) plug
   (d) pipe

5. What is a *tap drill?* _____
   _____
   _____

6. Using the table in Figure 9-11 in the text, list the proper tap drill sizes for the following threads.

   (a) 1/4-20UNC _____

   (b) 1/2-20UNF _____

   (c) 1/2-13UNC _____

   (d) 5/16-24UNF _____

   (e) 5/16-18UNC _____

   (f) 3/8-16UNC _____

7. Name the two types of tap wrenches. _____
   _____

8. How are broken taps removed from holes? _____

_____

_____

_____

_____

_____

_____

PART II

9. A die is used to cut _____ threads.   9._____

10. A(n) _____ holds the die and provides leverage for turning   10._____
the die on the work.

11. Ragged threads are the most common problem encountered when cutting external threads with a
die. List three causes for this problem.

_____

_____

_____

12. It has been established that the hole to be threaded must   12._____
be smaller than the thread. For example, a 5/16" diame-
ter hole must be drilled for a 3/8-16UNC thread. How
much larger must a shaft be machined to receive the
same size thread?

(a) 1/64" larger
(b) 1/32" larger
(c) 1/16" larger
(d) None of the above. The shaft must be _____.

13. List four safety practices that should be observed when hand threading.

1) _____

_____

2) _____

_____

3) _____

_____

4) _____

_____

Name_____

PART III

14. Secure a section of 3/8" diameter stock and cut a 3/8-16UNC thread on one inch of its length. Check finished thread size with a suitable nut.

15. Secure a suitable length and size piece of square stock. Use the appropriate tap drill for a 3/8-16UNC thread and drill one hole all of the way through the stock and another hole approximately 3/4 of the way through. Tap the first hole through. In the second, tap the thread to the bottom of the hole. Check for squareness and fit of threads with the previously threaded round stock.

FASTENERS

Name: _____ **Date:** _____

Instructor: _____ **Score:** _____

Carefully read the chapter. Then answer the following questions.

PART I

Match the definition with the correct term.

_____ 1. Threaded on both ends.

_____ 2. Used with a bolt.

_____ 3. Hammered into drilled or punched holes.

_____ 4. Used to position mating parts.

_____ 5. Fitted into a hole drilled crosswise in a shaft.

_____ 6. Available for internal and external applications; usually seated in a groove.

_____ 7. Used when parts are assembled permanently.

_____ 8. Provide an increased bearing surface for bolt heads and nuts.

_____ 9. Bonds surfaces together.

(a) Machine screw
(b) Rivet
(c) Nut
(d) Machine bolt
(e) Cotter pin
(f) Dowel pin
(g) Adhesive
(h) Stud bolt
(i) Drive screw
(j) Retaining ring
(k) Washer

10. Threaded fasteners use the _____ of the screw thread to clamp parts together.

10._____

11. When using various types of extractors, treatment with _____ will often make it easier to remove stubborn sheared bolts and machine screws.

11._____

12. A jam nut is used when _____.
    (a) two sections must be bolted together
    (b) a regular nut cannot be used
    (c) a regular nut must be locked in place
    (d) All of the above.

12._____

13. Castellated and slotted nuts have slots across the flats so they can be _____.
    (a) removed with a screwdriver
    (b) more easily cut with a chisel
    (c) locked in place with a cotter pin or safety wire
    (d) All of the above.

13._____

PART II

14. Where are wing nuts typically used? _____

_____

_____

15. What are *inserts?* When are they used? _____

_____

_____

_____

_____

16. Why are washers used? _____

_____

_____

_____

_____

17. Lock washers are used to _____.   17. _____
    (a) stop gears or pulleys from rotating on a shaft
    (b) prevent bolts and nuts from loosening under vibration
    (c) make a permanent joint
    (d) All of the above.

18. Name the three parts of the assembly shown below that are used to prevent a gear, wheel, or pulley from rotating on the shaft.

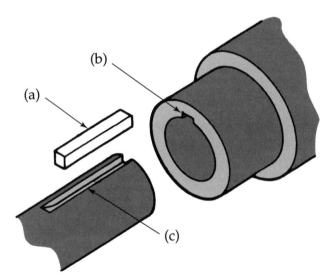

    (a) _____

    (b) _____

    (c) _____

Name_____

19. List the five steps involved in producing a solid adhesive bonded joint.

1) _____

_____

2) _____

_____

3) _____

_____

4) _____

_____

5) _____

_____

Identify the following fasteners.

20._____   21._____

22._____   23._____

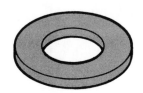

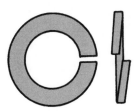

24._____   25._____

ART METAL

Name: _____   Date: _____

Instructor: _____   Score: _____

Carefully read the chapter. Then answer the following questions.

PART I

Match the definition with the correct term.

_____ 1. Heating and cooling process to soften metal.

_____ 2. Operation that removes oxide from metal.

_____ 3. Also called sculpturing in metal.

_____ 4. Using punches and light hammer to put designs in metal.

_____ 5. Used to raise sections of metal.

_____ 6. Process of giving three dimensional shape to flat sheet metal.

_____ 7. Used to support metal while it is being worked.

_____ 8. Process that smoothes metal with a special hammer.

_____ 9. Technique used to make shallow trays.

(a) Annealing
(b) Pickling
(c) Piercing
(d) Repoussé
(e) Chasing
(f) Raising
(g) Snarling iron
(h) Stake
(i) Beating down
(j) Planishing

10. _____, being almost pure tin, can be soldered so readily it is difficult to detect a properly made joint.

10. _____

11. Why is work bound together for soldering?_____

_____

_____

_____

12. How is copper annealed?_____

_____

_____

_____

13. When making a pickling solution _____.   13._____
    (a) the water is poured into the acid
    (b) the acid is poured into the water
    (c) neither acid or water is used
    (d) None of the above.

14. Piercing is done to _____.   14._____
    (a) place a smooth edge on holes and other openings
    (b) raise designs on the metal's surface
    (c) cut internal designs in metal
    (d) None of the above.

15. List four methods used for raising an object. _____

_____

_____

_____

_____

16. Sketch in the shapes of the hammer heads specified below.

Planishing   Forming

17. Sketch in the mallet heads specified below.

Double edge forming mallet   Round end forming mallet

Name_____

18. What size metal blank will be needed to make the circular bowl shown below?   18._____

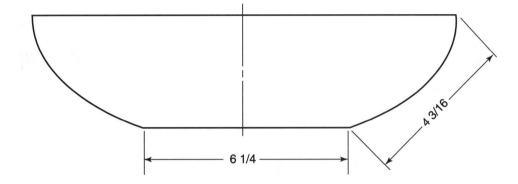

6 1/4

4 3/16

19. List five safety precautions that should be observed when working in art metal.

1) _____

_____

2) _____

_____

3) _____

_____

4) _____

_____

5) _____

_____

PART II

20. Using the space below and on the following page, design a mug, vase, bowl, wire, or strip sculpture. Get your ideas from nature, catalogs, craft magazines, etc.

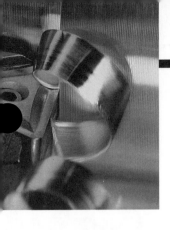

SHEET METAL

12

Name: _____   Date: _____

Instructor: _____   Score: _____

Carefully read the chapter. Then answer the following questions.

PART I

1. What is a sheet *metal pattern?* _____
   _____
   _____

2. What type of pattern development is used to make prisms and cylinders? _____
   _____

3. What type of pattern development is used to make cones and pyramids?_____
   _____

4. Of what use is a transition piece? Give an example. _____
   _____
   _____
   _____

Match the definition with the correct term.

_____ 5. Used to square and trim large metal sheets.

_____ 6. Makes small openings in the metal.

_____ 7. Circular work is cut with them.

_____ 8. Make it possible to join sheet metal sections.

_____ 9. Can make straight and circular cuts.

_____10. Gives additional strength and rigidity to sheet metal edges.

_____11. Designed to bend sheet metal to form edges and prepare the metal for a wire edge.

_____12. Enables the sheet metal worker to make a variety of bends by hand.

(a) Hems
(b) Wired edge
(c) Seams
(d) Circular snips
(e) Combination snips
(f) Hollow punch
(g) Squaring shears
(h) Metal stakes
(i) Bar folder

13. Make sketches of the following.

Single hem   Double hem   Wire edge

PART II

14. Because of the nature of sheet metal work, the observance of safe work procedures is most important. List five safety precautions that must be observed.

1) _____

_____

_____

_____

2) _____

_____

_____

_____

3) _____

_____

_____

_____

4) _____

_____

_____

_____

5) _____

_____

_____

_____

Name_____

PART III

15. Complete the stretchout of the prism.

Develop stretchout of a prism. Allow material for seams.

16. Complete the stretchout of the cylinder.

Develop stretchout of a cylinder. Allow material for seams. Assemble.

Name_____

17. Complete the stretchout of the pyramid.

Develop stretchout of a pyramid. Allow material for seams. Assemble.

18. Complete the stretchout of the cone.

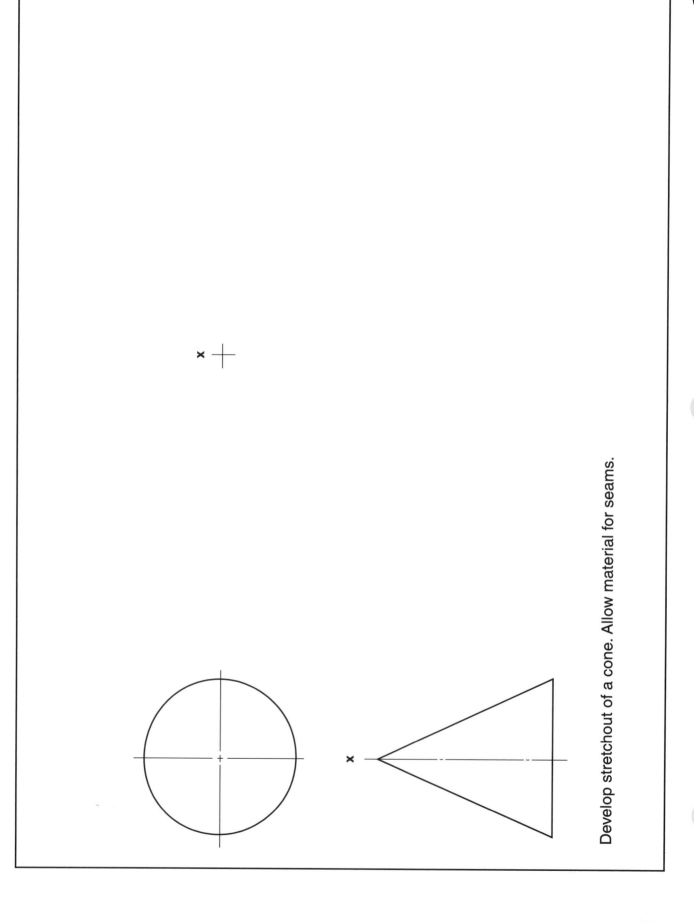

Develop stretchout of a cone. Allow material for seams.

Name_____

19. Complete the stretchout of the truncated prism.

3

2

1

4

3–4

1–2

1–2

Develop stretchout of a truncated prism. Allow material for seams. Assemble.

20. Make full size patterns for the following objects.

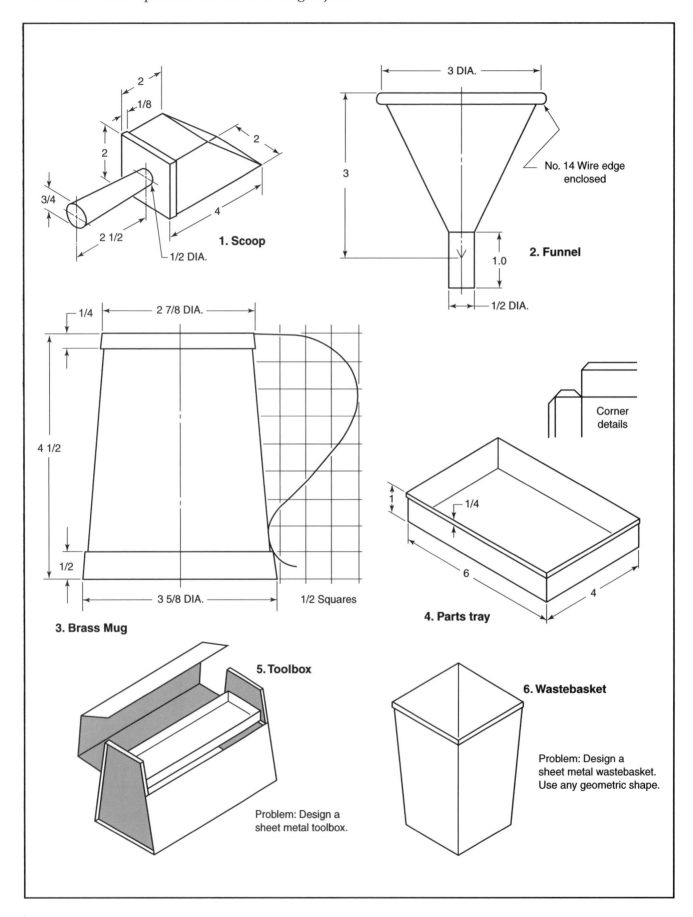

1. Scoop

2. Funnel

No. 14 Wire edge enclosed

3. Brass Mug

1/2 Squares

Corner details

4. Parts tray

5. Toolbox

Problem: Design a sheet metal toolbox.

6. Wastebasket

Problem: Design a sheet metal wastebasket. Use any geometric shape.

SOLDERING AND BRAZING

Name: _____   Date: _____

Instructor: _____   Score: _____

Carefully read the chapter. Then answer the following questions.

PART I

1. Define *soldering*. _____
   _____
   _____

2. Solders are alloys of _____ and _____.   2. _____
                                                                _____

3. What is the composition of 50-50 solder? _____
   _____

4. What is the composition of 60-40 solder? _____
   _____

5. Flux must be applied to joints being soldered to _____.   5. _____
   (a) remove existing oxides
   (b) prevent further oxides from forming
   (c) lower surface tension to permit solder to flow easier
   (d) All of the above.

6. What are the two categories of fluxes? _____
   _____

7. When a(n) _____ type of flux is used, the joint should be   7. _____
   cleaned with warm water.

8. Resin is a type of _____ flux that works best on tin plate   8. _____
   and brass.

9. List three ways of applying heat to a joint being soldered. _____
   _____
   _____
   _____

10. A soldering copper is tinned so molten solder will _____ to it.

10._____

11. How is a soldering copper tinned?_____

_____

_____

_____

_____

12. What are the three grades of hard or silver solder?_____

_____

_____

13. How does hard soldering differ from soft soldering? _____

_____

_____

14. What type of flame results from the combustion of perfect proportions of oxygen and the fuel gas?

_____

15. List five safety precautions that must be observed when soft and hard soldering.

1) _____

_____

2) _____

_____

3) _____

_____

4) _____

_____

5) _____

_____

PART II

Complete the following activities.

16. Tin a soldering copper.

17. Soft solder a lap joint.

18. Soft solder a seamed joint.

19. Sweat solder a joint.

20. Hard solder a simple lap joint.

SAND CASTING

Name: _____ **Date:** _____

Instructor: _____ **Score:** _____

Carefully read the chapter. Then answer the following questions.

PART I

1. What is a *mold?* _____
   _____
   _____

2. Molds made with moist sand are known as _____ molds. 2. _____

3. What is a match plate? _____
   _____
   _____
   _____

4. Molten metal reaches the opening or cavity through the _____ system. 4. _____

5. When is a core used in a mold? _____
   _____
   _____
   _____

6. How many parts are there to a simple pattern? 6. _____

7. How many parts are there to a split pattern? 7. _____

8. Why is a shrink rule used in place of a standard rule? _____
   _____
   _____

9. Why must draft be included in a pattern? _____
   _____
   _____
   _____

10. Name the parts of the two-part mold shown below.

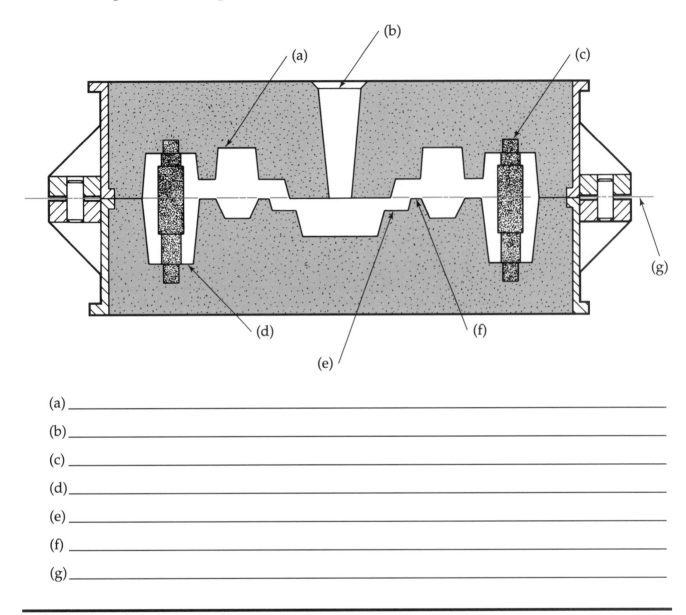

(a) _____

(b) _____

(c) _____

(d) _____

(e) _____

(f) _____

(g) _____

PART II

11. Complete the split pattern by drawing in the necessary draft.

Parting line

Name_____

12. Identify the foundry tools.

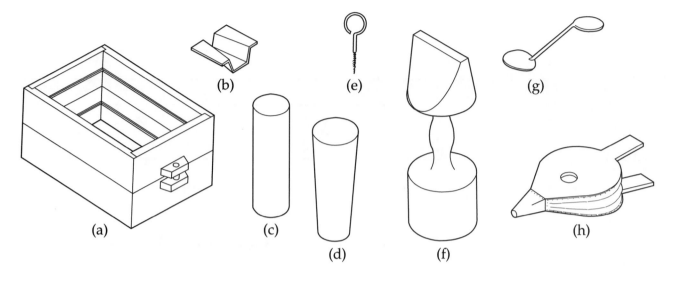

(a) _____

(b) _____

(c) _____

(d) _____

(e) _____

(f) _____

(g) _____

(h) _____

13. When making a sand mold, the sand must be broken down into fine loose particles with a(n) _____.

13. _____

14. Sand is packed around the pattern by hand with a(n) _____.

14. _____

15. Casting metal is usually melted in a(n) _____.

15. _____

16. Molten metal temperature is usually checked with a(n) _____.

16. _____

17. List five safety precautions that must be observed when sand casting.

1) _____

_____

2) _____

_____

3) _____

_____

4) _____

_____

5) _____

_____

PART III

18. Make a mold using a simple pattern.

19. Make a mold using a split pattern.

20. Make a mold in which a core must be used.

METAL CASTING TECHNIQUES

Name: _____ Date: _____

Instructor: _____ Score: _____

Carefully read the chapter. Then answer the following questions.

PART I

1. Describe the die casting process. _____

   _____

   _____

   _____

2. Die casting molds (dies) are made from _____.   2._____
   - (a) aluminum
   - (b) special sand
   - (c) hardened steel
   - (d) None of the above.
   - (e) All of the above.

3. Die casting metals are usually alloys of which metals? _____

   _____

   _____

   _____

4. How does permanent mold casting differ from castings made using sand molds? _____

   _____

   _____

   _____

5. How is a centrifugal casting made? _____

   _____

   _____

   _____

   _____

   _____

   _____

6. When slush molding, molten metal is poured into the   6._____
   mold and left _____.
   (a) to form a thick shell of metal in the mold
   (b) in to make a solid mold
   (c) long enough to form a thin shell of metal in the mold
   (d) None of the above.

7. Briefly describe the investment casting process. _____

   _____

   _____

   _____

8. What type of materials are used to make investment casting patterns? _____

   _____

   _____

9. Shell molding is a foundry process in which the molds   9._____
   are made in the form of _____.

10. What is unique about the lost foam casting process? _____

    _____

    _____

    _____

    _____

11. Why are plaster molds fired or baked at low temperature to remove all traces of moisture?

    _____

    _____

    _____

12. Describe the laminated object manufacturing process (LOM). _____

    _____

    _____

    _____

    _____

    _____

13. What is unique about the stereolithography process? _____

    _____

    _____

    _____

    _____

Name_____

14. List three safety precautions that must be observed when performing casting operations.

1) _____

_____

_____

_____

_____

2) _____

_____

_____

_____

_____

3) _____

_____

_____

_____

_____

PART II

15. Secure samples of the following: 15._____

   (a) Die castings.
   (b) Permanent mold casting.
   (c) Slush mold casting.
   (d) Investment casting pattern.
   (e) Lost foam pattern.
   (f) Object made by one of the rapid prototyping
       processes.

WROUGHT METAL

Name: _____   Date: _____

Instructor: _____   Score: _____

Carefully read the chapter. Then answer the following questions.

PART I

1. Originally, wrought ironwork consisted of metal prod-   1._____
   ucts formed and shaped by a(n) _____.

2. Name and explain two factors that must be considered when making bends in metal.

   1) _____

   _____

   _____

   _____

   2) _____

   _____

   _____

   _____

3. _____ angles are made by squeezing the metal in a vise   3._____
   after the initial bend has been made.

4. Why should a full-size pattern of a desired scroll or other curved section be made before attempting to form the metal?

   _____

   _____

   _____

5. Curved sections can be made _____.   5._____
   (a) over the anvil horn
   (b) using a bending jig
   (c) using a bending fork
   (d) All of the above.
   (e) None of the above.

6. What is a *metal former?* _____

_____

_____

7. What are two methods often used to assemble wrought iron work? _____

_____

_____

8. List five safety precautions that must be observed when working on wrought iron projects.

1) _____

_____

2) _____

_____

3) _____

_____

4) _____

_____

5) _____

_____

PART II

9. Draw the double end scroll shown in Figure 16-14 of the text.

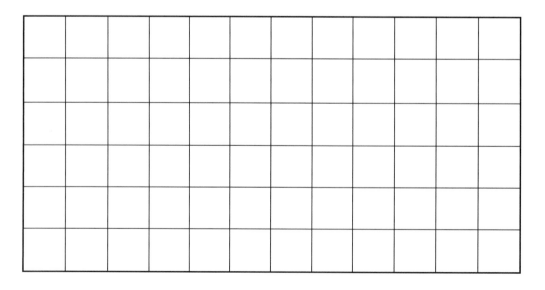

Name_____

10. Draw the scroll shown in Figure 16-15 in the text.

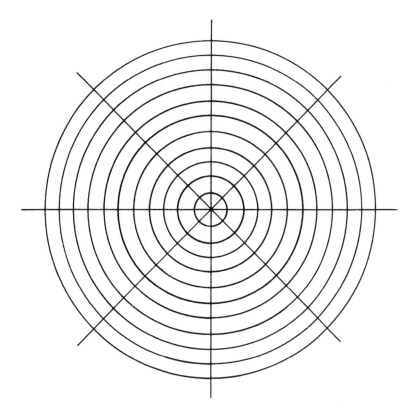

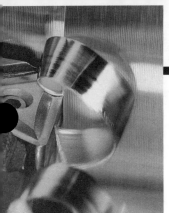

FORGING

Name: _____   Date: _____

Instructor: _____   Score: _____

Carefully read the chapter. Then answer the following questions.

PART I

1. Define *forging*. _____

   _____

   _____

   _____

2. Forging is one of the few metalworking processes that improves the _____ of most metals.

   2. _____

3. Hot metal is held in _____ while being formed.

   3. _____

4. Identify the parts of the anvil.

   (a) _____

   (b) _____

   (c) _____

   (d) _____

   (e) _____

   (f) _____

(a)   (b)

(c)

(d)

(e)

(f)

5. Identify the hand forging tools shown below.

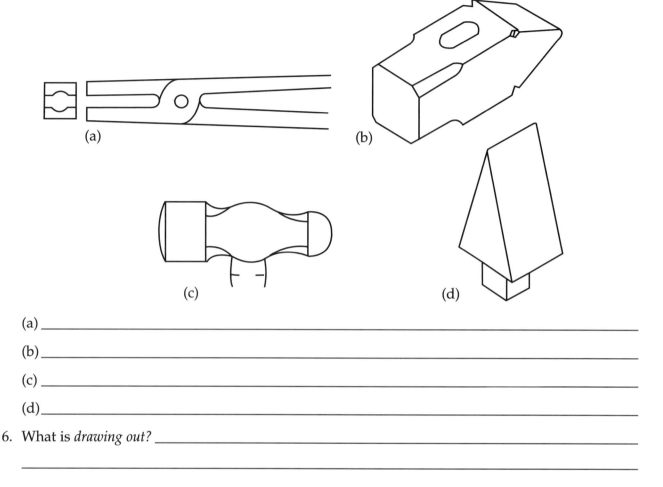

(a)

(b)

(c)

(d)

(a) _____

(b) _____

(c) _____

(d) _____

6. What is *drawing out?* _____

_____

_____

7. What is *upsetting?* _____

_____

_____

Match each definition with the correct term.

_____ 8. Hardened steel blocks with cavities that shape the forging.

_____ 9. Work is manipulated by hand with this forging technique.

_____ 10. Shaped dies replace the flat dies of open die forging.

_____ 11. Surplus metal that forms at the parting line of dies.

_____ 12. Metal is shaped by gradual application of pressure.

_____ 13. Use to reduce or taper the diameter or thickness of a bar.

_____ 14. Forging technique in which metal is shaped by a series of rapid hammerlike blows.

(a) Open die forging
(b) Drop forge
(c) Precision forging
(d) Dies
(e) Roll forging
(f) Cold forming
(g) Intraform
(h) Flash

Name_____

PART II

15. Draw out a piece of 1/2″ (12.5 mm) diameter metal to a point.

16. Secure a section of 1/2″ (12.5 mm) diameter metal that is 4″ (100 mm) long and upset it until its length is 2 1/2″ (62.5 mm) long. What happens to the metal when it is overheated?

_____

_____

_____

_____

_____

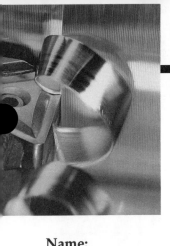

HEAT TREATMENT OF METALS

18

Name: _____ Date: _____

Instructor: _____ Score: _____

Carefully read the chapter. Then answer the following questions.

PART I

1. Heat treating operations involve the _____
   _____
   _____
   _____
   _____
   _____.

2. _____ steel cannot be hardened by conventional heat   2. _____
   treating techniques. However, a hard shell can be put on   _____
   its surface by the process called _____.

3. List two reasons an electric heat treating furnace is preferred over a gas-fired furnace. _____
   _____
   _____
   _____
   _____

Name the heat treating processes described in questions 2 through 12.

4. Technique used to soften metals and make them easier to   4. _____
   machine.

5. Cooling rapidly in water, brine, oil, liquid nitrogen, or   5. _____
   blasts of cold air.

6. Reduces stress that has developed in parts that have been   6. _____
   cold worked, machined, or welded.

7. Only the outer surface of the metal is hardened.   7. _____

8. Technique used to obtain optimum physical qualities in   8. _____
   steel. Metal may become hard and brittle.

9. Operation that reduces hardness and brittleness of the above steel.

9._____

10. Measures and controls furnace temperature.

10._____

11. Used to check whether metal is in required condition of heat treatment.

11._____

12. Hardness tester that uses a steel ball that is forced into the material.

12._____

13. Hardness tester that uses a diamond penetrator to measure the degree of hardness.

13._____

14. Portable hardness tester that does not mar the surface of the material being tested.

14._____

15. List five safety precautions that must be observed when heat treating metal.

1) _____

_____

2) _____

_____

3) _____

_____

4) _____

_____

5) _____

_____

PART II

16. Properly heat treat a center punch.

17. Demonstrate the proper way to case harden a section of mild steel. Use Kasenit or similar safe commercial type casehardening compound.

GAS WELDING

Name: _____ Date: _____

Instructor: _____ Score: _____

Carefully read the chapter. Then answer the following questions.

PART I

1. Describe the welding process. _____

_____

_____

_____

_____

_____

_____

2. Gas welding includes a group of welding processes that _____

_____

_____

_____.

3. Basically, oxyacetylene welding equipment consists of the following items. Describe why or how they are used.

(a) Cylinders: _____

_____

(b) Regulators: _____

_____

(c) Hoses: _____

_____

(d) Torch: _____

_____

4. The _____ hose is colored green. The _____ hose is colored red.

    4._____

    _____

5. What type of device should be used to light a welding torch?

5._____

6. Welding rod is often used when gas welding to build up the _____ and make it as strong as the _____.

6._____

_____

7. Why is bronze filler rod preferred to brass when brazing or bronze welding? _____

_____

_____

8. List the three reasons flux must be used when gas welding most nonferrous metals and cast iron.

_____

_____

_____

9. An excess of acetylene produces a(n) _____ flame.

9._____

10. A sharp hissing sound indicates that there is an excess of _____, and a(n) _____ flame is the result.

10._____

_____

11. Draw and label the five basic types of weld joints.

Name_____

12. The _____ welding technique is used for joining thin   12._____
    metal. In which direction is the flame directed?

    _____

    _____

    _____

13. Heavier sections are joined using the _____ welding   13._____
    technique. In which direction is the flame directed?

    _____

    _____

    _____

14. List five safety precautions that must be observed when gas welding.

    1) _____

    _____

    2) _____

    _____

    3) _____

    _____

    4) _____

    _____

    5) _____

    _____

PART II

15. Practice making the various joints with an acetylene torch.

SHIELDED METAL ARC WELDING

20

Name: _____ Date: _____

Instructor: _____ Score: _____

Carefully read the chapter. Then answer the following questions.

PART I

1. What is *shielded metal arc welding?* _____

_____

_____

_____

_____

2. What is the difference between dc and ac current? _____

_____

_____

_____

_____

3. How are arc welders rated? _____

_____

_____

_____

4. What are arc welding electrodes? _____

_____

_____

_____

5. What purpose do electrodes serve? _____

_____

_____

_____

6. Of what use is the flux on arc welding electrodes? _____

_____

_____

_____

_____

7. What purpose does the ground clamp serve? _____

_____

_____

_____

8. Under what type of connection is the circuit considered to be DCEP? _____

_____

_____

_____

_____

9. Under what type of connection is the circuit considered to be DCEN? _____

_____

_____

_____

_____

10. Draw in the correct weld symbols for the welds shown.

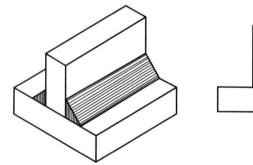 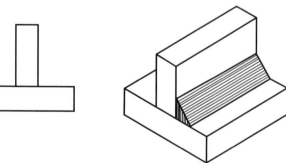 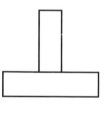

Name_____

11. Draw in the correct welds for the symbols shown.

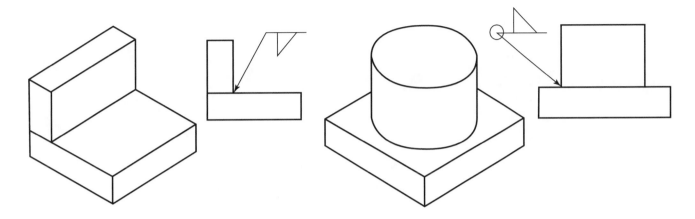

12. Complete the drawing sketching the degrees of peneteration in each situation.

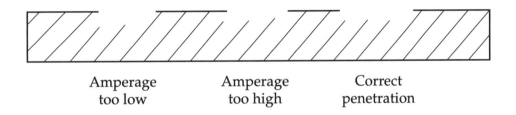

<table>
<tr><td>Amperage
too low</td><td>Amperage
too high</td><td>Correct
penetration</td></tr>
</table>

13. List five safety precautions that must be observed when arc welding.

1) _____

_____

2) _____

_____

3) _____

_____

4) _____

_____

5) _____

_____

PART II

14. Practice running beads until you can do so satisfactorily.

15. Prepare and weld the five basic weld joints.

OTHER WELDING PROCESSES

Name: _____ **Date:** _____

Instructor: _____ **Score:** _____

Carefully read the chapter. Then name the welding process described in each of the following.

1. A gas shielded arc welding technique that uses a permanent electrode (the electrode is not consumed).

1._____

2. A gas shielded arc welding technique in which the electrode melts and contributes filler metal to the joint.

2._____

3. A type of arc welding where the consumable metal electrode is shielded by a blanket of flux that covers the weld area.

3._____

4. Fusion is produced by an electric arc between the metal stud, or similar part, and the work surface.

4._____

5. A group of welding techniques that use pressure, electric current, electrical resistance of the work, and the resulting heat to join metal sections.

5._____

6. The best known of the welding process described in statement 5. It is widely used because it saves time and weight, and is adaptable to robotic techniques.

6._____

7. Makes use of fast moving electrons to supply the energy to melt and fuse the parts being joined.

7._____

8. Uses frictional heat and pressure to produce full strength welds in a matter of seconds.

8._____

9. A process for joining metals without using solder, fluxes, or filler metals; usually does not require the application of external heat.

9._____

10. Welding technique used to attach leads to microcircuits.

10._____

11. Pressure alone is used to join two metals together. Requires special tools.

11._____

12. The welding process that uses radiation to generate the heat required for metal fusion.

12._____

13. Wire is fed through a special spray gun, melted by a gas flame, and atomized and sprayed onto the work by compressed air.

13._____

14. Involves the application of metals, and other materials, that cannot be drawn into wire. They are used in powder form.

14._____

15. The process uses an electric arc that is contained in a water-cooled jacket. Any inorganic materials that can be melted without decomposition at temperatures of 30,000°F (16,650°C) can be applied.

15._____

METAL FINISHES 22

Name: _____ Date: _____

Instructor: _____ Score: _____

Carefully read the chapter. Then answer the following questions.

PART I

1. What is an *abrasive?* _____
_____
_____
_____
_____

2. The performance and efficiency of an abrasive is determined by_____
_____
_____
_____
_____.

3. List four reasons for which finishes are applied to metals. _____
_____
_____
_____
_____

4. _____ is a natural abrasive that is black in color and cuts slowly.

4._____

5. Which manufactured abrasive is used when much metal must be removed?

5._____

6. _____ is bright red in color and used for cleaning and polishing when a minimum of metal is to be removed.

6._____

7. List three safety precautions that must be observed when hand polishing with abrasives.

   1) _____

   _____

   2) _____

   _____

   3) _____

   _____

8. _____ is accomplished by spraying the metal surface with a fine abrasive driven by compressed air.

   8. _____

9. _____ is used to produce a smooth satin sheen on the metal.

   9. _____

10. The application in which a coating is applied by dipping the steel in molten zinc is called _____.

    10. _____

11. What process uses the action of an electric current to deposit a coating of metal on the surface of another metal?

    11. _____

12. In flame spraying, metal is heated to its _____ point and sprayed by air pressure onto the work.

    12. _____

13. How do organic coatings (with the exception of the epoxies) set? _____

    _____

    _____

    _____.

14. Epoxies require the addition of a(n) _____ to set.

    14. _____

15. List the methods used to apply organic coatings. _____

    _____

    _____

    _____

    _____

16. _____ is an electrochemical process used to thicken the oxide film on aluminum. Color may also be added.

    16. _____

17. _____ improves the surface smoothness and reflectance of aluminum by an electrochemical process.

    17. _____

18. Which process produces a black oxide coating on the machined surfaces of ferrous metals?

    18. _____

Name_____

19. List five safety precautions that must be observed when using and applying finishes to metals.

1) _____

_____

2) _____

_____

3) _____

_____

4) _____

_____

5) _____

_____

PART II

20. Secure samples of various abrasives used for hand polishing. Attach each to a space below and identify it.

GRINDING

Name: _____   **Date:** _____

Instructor: _____   **Score:** _____

Carefully read the chapter. Then answer the following questions.

PART I

1. How does grinding remove material? _____

   _____

   _____

2. Grinding is used for _____.   2._____
   (a) sharpening tools
   (b) removing material too hard to be machined by other
       techniques
   (c) making fine surface finishes and meeting close
       tolerances
   (d) All of the above.

3. To prevent work being ground on a bench or pedestal   3._____
   grinder from being wedged between the wheel and the
   tool rest, it is recommended that the rest be adjusted

   _____.
   (a) to within 1/8″ (3.0 mm) of the wheel
   (b) as close to the wheel as possible
   (c) to within 1/16″ (1.5 mm) of the wheel
   (d) None of the above.

4. How can a grinding wheel be checked for soundness? _____

   _____

   _____

   _____

5. A wheel dresser is used on a grinding wheel to _____.   5._____
   (a) test the soundness of the wheel
   (b) true the wheel and remove the glaze
   (c) adjust the spacing between the tool rest and the
       wheel
   (d) All of the above.

6. List three recommendations that will secure maximum efficiency from a bench or pedestal grinder.

   1) _____

   _____

   2) _____

   _____

   3) _____

   _____

7. Precision grinding is an economical way to _____

   _____

   _____

   _____.

Match the definition with the correct term.

_____ 8. The grinding wheel is shaped to produce the required design on the work.

_____ 9. Work is mounted between centers and rotates while in contact with the grinding wheel.

_____10. Work not supported by centers as it rotates against the grinding wheel.

_____11. Technique used to grind flat surfaces.

_____12. Done to produce a fine surface finish accuracy on internal diameters.

(a) Cylindrical grinding

(b) Centerless grinding

(c) Internal grinding

(d) Tool and cutter grinding

(e) Form grinding

(f) Surface grinding

PART II

13. The _____ surface grinder uses a reciprocating motion to move the worktable back and forth under the grinding wheel.

    13. _____

14. Why is it sometimes necessary to demagnetize work held on a magnetic chuck? _____

    _____

    _____

    _____

15. _____ grinding permits machining of work that is wider than the face of the grinding wheel.

    15. _____

Name_____

16. The ideal grinding wheel would be one in which _____

_____

_____

_____

_____.

17. _____ and _____ will prevent a grinding wheel from   17._____
    cutting freely.
    _____

18. A grinding wheel 32A46-H8V is specified for a grinding job. What do the series of numbers and letters mean?

    (a) 32A: _____

    _____

    (b) 46: _____

    _____

    (c) H: _____

    _____

    (d) 8: _____

    _____

    (e) V: _____

    _____

19. List five safety precautions that must be observed when grinding.

    1) _____

    _____

    2) _____

    _____

    3) _____

    _____

    4) _____

    _____

    5) _____

    _____

PART III

20. Using this page or a blank (unlined) sheet of paper, prepare a safety poster for bench or pedestal grinding.

DRILLS AND DRILLING MACHINES

Name: _____ **Date:** _____

Instructor: _____ **Score:** _____

Carefully read the chapter. Then answer the following questions.

PART I

1. How is drill press size determined? _____
   _____
   _____
   _____

2. Common drills are often called _____.

   2. _____

3. A(n) _____ drill has coolant holes through the body that permit fluid or air to be forced to the point to remove heat from the point.

   3. _____

4. _____ are used to drill holes smaller than 0.0135″ in diameter.

   4. _____

5. What is unique about indexable-insert drills? _____
   _____
   _____
   _____

6. Fill in the name of the drill series described.

   (a) #80 to #1 _____

   (b) A to Z _____

   (c) 1/64″ to 3 1/2″ _____

   (d) 3.0 mm to 76.0 mm _____

7. The size of a frequently used drill is often worn from the tool's shank. You can determine the drill's size by measuring it with a(n) _____ or _____.

   7. _____
   _____

Match each definition with the correct term.

_____ 8. The amount of surface of the point is relieved back from the tips.

_____ 9. The portion of the point, back from the lips or cutting edges.

_____10. The sharp edge at the extreme tip of the drill.

_____11. Used to hold straight shank drills.

_____12. The cone-shaped end of the drill that does the cutting.

_____13. The metal column that separates the drill flutes.

_____14. Found on the end of a taper shank drill.

_____15. Must be mounted in a chuck.

_____16. The cutting edges of the drill.

_____17. Fits directly in the drill press spindle.

_____18. The narrow strip extending back the entire length of each flute.

_____19. Two or more spiral grooves that run the entire length of the drill body.

(a) Drill point
(b) Dead center
(c) Lips
(d) Heel
(e) Lip clearance
(f) Straight shank
(g) Taper shank
(h) Flutes
(i) Tang
(j) Margin
(k) Web
(l) Chuck

20. Identify the three drilling accessories shown below.

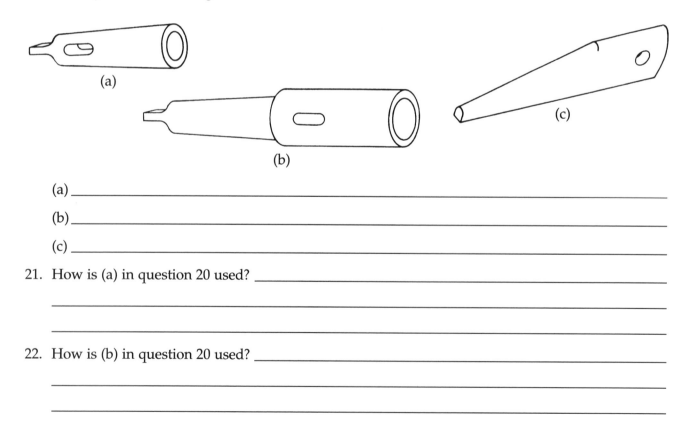

(a)
(b)
(c)

(a) _____

(b) _____

(c) _____

21. How is (a) in question 20 used? _____

_____

_____

22. How is (b) in question 20 used? _____

_____

_____

Name_____

23. How is (c) in question 20 used? _____

_____

_____

_____

24. Why is it important for work to be clamped solidly to the drilling machine? _____

_____

_____

_____

_____

25. Why are cutting fluids used when drilling? _____

_____

_____

_____

26. When repointing or sharpening a drill, three factors must be considered. List them. _____

_____

_____

_____

_____

27. The tool used to cut a chamfer in a drilled hole so a flat head screw can be fitted is called a(n) _____.

27._____

28. The _____ is employed to prepare a drilled hole to receive a fillister head or socket head screw.

28._____

29. The operation that machines a circular spot around a hole drilled on a rough surface so a bolt head or washer and nut can be seated properly is called _____.

29._____

30. A reamed hole _____ a drilled hole.
    (a) is not as accurate as
    (b) has a better finish and is more accruate than
    (c) is just another name for
    (d) All of the above.

30._____

31. A reamed hole must first be drilled slightly _____ than the required hole size.

31._____

32. List four safety practices that must be observed when using drills and drilling machines.

1) _____

_____

2) _____

_____

3) _____

_____

4) _____

_____

PART II

Solve the following problems using the information in the drill speed and feed table (text Figure 24-36) and the formula for determining rpm. Use the mid range of the cutting speeds shown. For example, if the recommended range of cutting speeds is 200–300 fpm, the mid range would be 250 fpm. Round your answers.

The simple formula $rpm = \dfrac{(4 \times CS)}{D}$ will determine the rpm to operate any diameter drill (D) at any specified cutting speed.

33. At what rpm must a drill press operate when drilling aluminum with a 1/8″ diameter drill?

34. At what rpm must a drill press operate when drilling free machining brass with a 1/2″ diameter drill?

35. At what rpm must a drill press operate when drilling free machining steel with a 3/8″ diameter drill?

Name_____

Metric problems are solved in a similar manner using the following formula:

$$rpm = \frac{CS \times 1000}{D \times \pi}$$

Where: CS   = Cutting speed (mpm)
           D   = Drill diameter (mm)
           π = 3 (rounded)

36. At what rpm must a drill press operate when drilling copper with a 13.5 mm diameter drill?

37. At what rpm must a drill press operate when drilling aluminum with a 20 mm diameter drill?

PART III

38. Demonstrate how a center finder or wiggler is used to align work for drilling.

39. Sharpen a drill that has become dull. Test it after sharpening by drilling a hole and compare the drilled size against the size of the drill.

40. Drill a hole that is centered across the diameter of a section of round stock.

SAWING AND CUTOFF MACHINES

Name: _____ **Date:** _____

Instructor: _____ **Score:** _____

Carefully read the chapter. Then answer the following questions.

PART I

1. List the three types of metal cutting power saws. Briefly describe how each type operates.

   1) _____

   _____

   _____

   _____

   2) _____

   _____

   _____

   _____

   3) _____

   _____

   _____

   _____

2. Proper blade selection is important. In general, large sections and soft materials require a(n) _____ blade. Small sections or thin work and hard materials require a(n) _____ blade.

   2. _____
   _____

3. What is meant by the three-tooth rule? _____

   _____

   _____

4. _____ blades are used where safety requirements demand a shatterproof blade.

   4. _____

5. _____ blades are the first choice for straight, accurate cutting under a variety of conditions.

   5. _____

6. If long life and accurate cuts are to be achieved, the blade   6. _____
   must be properly _____.

7. What type of cutting action does a band saw use to cut metal? _____

   _____

   _____

   _____

   _____

8. What are three advantages the band saw offers over other types of power saws? _____

   _____

   _____

   _____

9. Explain two methods of getting the proper tension on the blade of a reciprocating type saw.

   1) _____

   _____

   _____

   _____

   2) _____

   _____

   _____

   _____

10. Identify the blades shown below.

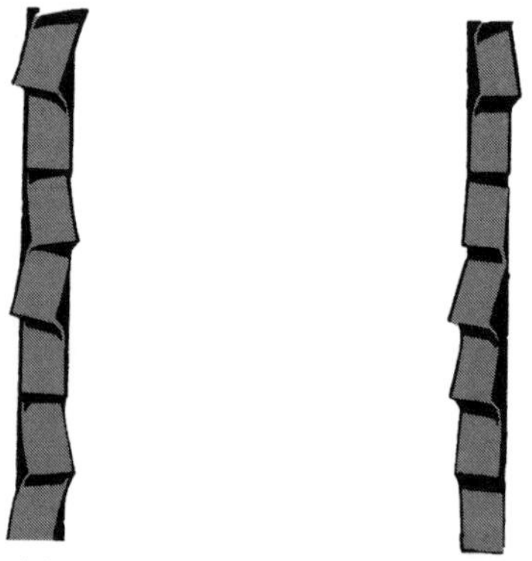

                        (a)   (b)

   (a) _____

   (b) _____

Name_____

11. Sketch in and label the tooth shapes indicated.

   (a) Standard tooth

   (b) Skip tooth

   (c) Hook tooth

12. What problems cause the following?

   (a) Broken blades: _____

   _____

   _____

   _____

   (b) Crooked cuts: _____

   _____

   _____

   _____

   (c) Teeth strip off: _____

   _____

   _____

   _____

13. List the three types of circular cutoff saws. _____

   _____

   _____

14. Which type of circular cutoff saw may or may not have teeth? If teeth are on the blade, why are they used?

_____

_____

_____

_____

15. List five safety precautions that must be observed when using a power saw.

1) _____

_____

2) _____

_____

3) _____

_____

4) _____

_____

5) _____

_____

METAL LATHE

26

Name: _____ **Date:** _____

Instructor: _____ **Score:** _____

Carefully read the chapter. Then answer the following questions.

PART I

Match the definition with the correct term.

_____ 1. Contains the dead center and supports the outer end of the work.

_____ 2. Contains the spindle.

_____ 3. Various work-holding attachments are fitted to it.

_____ 4. Foundation of the lathe.

_____ 5. Engaged for thread cutting.

_____ 6. Controls the operation of power feeds.

_____ 7. Indicates how to position levers for cutting various pitch threads and feed combinations.

_____ 8. Contains the saddle and cross slide.

_____ 9. Transmits power to the carriage through a gearing and clutch arrangement.

_____ 10. Regulates the distance of tool travel per spindle revolution.

_____ 11. Used to obtain slower spindle speeds on belt driven lathes.

_____ 12. Transmits power through a chain of gears.

(a) Lathe bed
(b) Headstock
(c) Spindle
(d) Back gears
(e) Tailstock
(f) Carriage
(g) Lead screw
(h) Quick-change gearbox
(i) Feed mechanism
(j) Feed change levers
(k) Index plate
(l) Half-nut

13. _____ cuts are the deep cuts made to remove considerable metal from a workpiece.

13. _____

14. _____ tools are ground to prevent interference with the tailstock center.

14. _____

15. Lathe size is determined by its swing and bed length.   15. _____
    Swing is the largest _____ work that can be turned. Bed
    length is the entire length of the _____.   _____

16. Each of the lathe parts fall into one of three functional divisions. List them. _____

    _____

    _____

    _____

17. A(n) _____ is used to remove taper shank tools from the   17. _____
    headstock spindle.

PART II

Calculate the correct rpm for machining the following materials. Round your answer off to the nearest 50 rpm. Do your calculations in the space provided. Follow the formula, $rpm = \dfrac{CS \times D}{4}$ *where:*

    rpm = revolutions per minute
    CS   = cutting speed of the particular metal being machined in feet per minute (fpm).
    D   = diameter of work in inches

18. Mild steel—3 1/2″ in diameter

19. Aluminum—1.250″ in diameter

20. Tool steel—4.5″ in diameter

Name_____

PART III

21. What problems will occur with work turned between centers when the live center does not run true?

_____

_____

22. Name the type of lathe chucks described below.

   (a) Each jaw operates individually, permitting irregularly shaped work to be centered. Jaws can be reversed.

   _____

   (b) Jaws operate simultaneously and round stock is centered automatically. Jaws cannot be reversed.

   _____

   (c) Operates like a universal chuck. Is useful for holding small work for turning and (when fitted in the tailstock) for drilling.

   _____

23. Explain how a lathe is prepared for operation. _____

_____

_____

_____

_____

_____

_____

_____

24. List five safety precautions that must be observed when operating a lathe. _____

   1) _____

   _____

   2) _____

   _____

   3) _____

   _____

   4) _____

   _____

   5) _____

   _____

25. Describe the parting operation. _____

_____

_____

_____

_____

_____

CUTTING TAPERS AND SCREW THREADS ON A LATHE

Name: _____   **Date:** _____

Instructor: _____   **Score:** _____

Carefully read the chapter. Then answer the following questions.

PART I

1. There are four generally accepted methods for machining tapers on the lathe. List the advantage or disadvantage of using each of the following methods.

   (a) Compound rest _____

   _____

   (b) Offset tailstock_____

   _____

   (c) Taper attachment_____

   _____

   (d) Cutting tool ground to desired angle _____

   _____

2. What is *setover?* _____

   _____

   _____

PART II

The following terms are used when calculating tailstock setover:

   D = Diameter at large end of work
   d = Diameter at small end of work
   l = Length of taper
   L = Total length of work

Use the following formulas for calculating setover.

When taper per inch is given

$$\text{Offset} = \frac{L \times TPI}{2}$$

   TPI = Taper per inch
   L   = Total length of work

When taper per foot is given

When the *taper per foot* is known, it must be converted to *taper per inch*. The following formula takes this into account:

$$\text{Offset} = \frac{\text{TPF} \times \text{L}}{24}$$

When taper per millimeter is given

$$\text{Offset} = \frac{\text{L (in mm)} \times \text{TPmm}}{2}$$

When dimensions of tapered section are given but TPF, TPI, or TPmm are not given

Quite often plans do not specify TPF, TPI, or TPmm but do give other pertinent information. Calculations will be easier if all fractions are converted to decimals. All dimensions must be in either inches or millimeters.

$$\text{Offset} = \frac{\text{L} \times (\text{D} - \text{d})}{2 \times l}$$

Calculate the tailstock setover for the following problems.

3. Taper per inch = 0.031″

    Length of piece = 6.00″

4. Taper per foot = 0.185″

    Length of piece = 13.000″

5. Large diameter = 1.750″

    Small diameter = 1.500″

    Length of taper = 6.000″

    Length of piece = 10.000″

Name_____

6. Taper per mm = 0.007 mm

   Length of piece = 1750.0 mm

7. Large diameter = 55.0 mm

   Small diameter = 45.0 mm

   Length of taper = 350.0 mm

   Length of piece = 565.0 mm

PART III

8. Name five applications for screw threads. _____

_____

_____

_____

Match the following.

_____ 9. The largest diameter of the thread.

_____ 10. The distance from one point on a thread to the correspond-
ing point on the next thread.

_____ 11. The smallest diameter of the thread.

_____ 12. The distance a nut or threaded section will travel in one full
revolution of the screw.

_____ 13. Diameter of an imaginary cylinder that would pass through
threads at such points to make the width of thread and the
width of the spaces at these points equal.

(a) Pitch diameter

(b) Pitch

(c) Root diameter

(d) Lead

(e) Major diameter

14. The _____ groove provides a place to stop the threading
tool at the end of its cut.

14._____

15. A(n) _____ is used to indicate when to engage the half-nuts
to permit the cutting tool to follow in the original cut.

15._____

Using the 3-wire method for measuring screw threads, calculate the correct measurement over the wires for the following threads. Use the wire size given. Do your calculations in the space provided. Follow the formula, $M = D + 3G - \dfrac{1.5155}{N}$ where:

    M = Measurement over the wires

    D = Major diameter of thread

    G = Diameter of wires

    N = Number of threads per inch

16. 7/16 – 20UNF (wire size 0.032″)

17. 1/4 – 28UNF (wire size 0.021″)

18. 3/4 – 10UNC (wire size 0.058″)

19. List five safety precautions that must be observed when operating a lathe.

    1) _____

    _____

    2) _____

    _____

    3) _____

    _____

    4) _____

    _____

    5) _____

    _____

OTHER LATHE OPERATIONS

Name: _____   Date: _____

Instructor: _____   Score: _____

Carefully read the chapter. Then answer the following questions.

1. Besides turning work on the lathe, many other machine operations can be performed on the machine tool. List four such machining operations. _____

   _____

   _____

   _____

   _____

2. When reaming on a lathe, you should use a cutting speed about _____ the speed you would use for a similar size drill with the material being machined.

   (a) one-half
   (b) two-thirds
   (c) three-fourths
   (d) five-sixths

   2. _____

3. The main purpose of boring is to produce a hole that is concentric with the _____ (inside/outside) diameter of the work.

   3. _____

4. What will happen if the knurling tool setup is *not* made properly? _____

   _____

   _____

   _____

   _____

5. Why is the left-hand method of filing preferred over the right-hand method? _____

   _____

   _____

   _____

   _____

   _____

6. Why is it important to clean the lathe thoroughly after polishing operations? _____

_____

_____

_____

_____

7. A(n) _____ is a cylindrical piece of hardened steel that has been machined with a very slight taper.

7. _____

8. Work is pressed on the above tool with a(n) _____.

8. _____

9. What is a *turret lathe?* _____

_____

_____

_____

10. List five safety precautions that must be observed when operating a lathe.

1) _____

_____

2) _____

_____

3) _____

_____

4) _____

_____

5) _____

_____

BROACHING OPERATIONS

Name: _____   Date: _____

Instructor: _____   Score: _____

Carefully read the chapter. Then answer the following questions.

PART I

1. Broaching is a manufacturing process for machining _____.   1. _____

   (a) helical splines
   (b) flat, rounded, and contoured surfaces
   (c) internal and external surfaces
   (d) All of the above.
   (e) None of the above.

2. Why does internal broaching require a starting hole? _____

   _____

   _____

   _____

   _____

3. Internal broaching makes use of a(n) _____ broach.   3. _____

4. The _____ broach is used for external broaching.   4. _____

5. In _____ broaching, the tool is stationary and the work is   5. _____
   pulled through or over it.

6. What are the three kinds of teeth on a broaching tool? _____

   _____

   _____

   _____

7. Broaching tools have long lives because _____

   _____

   _____

   _____

   _____.

8. The broaching operation _____.

    (a) is usually completed in a single pass of the cutting tool

    (b) can remove metal faster in one pass than any other machining technique

    (c) can maintain consistently close tolerances

    (d) All of the above.

    (e) None of the above.

8._____

9. The broaching operation offers several advantages including _____.

    (a) long tool life

    (b) maintains close tolerances

    (c) produces good surface finishes

    (d) All of the above.

    (e) None of the above.

9._____

10. The surface finish of a broached surface can further be improved by adding _____ to the finishing end of the tool.

10._____

PART II

11. Cut a keyway in a gear or pulley.

MILLING
MACHINES

Name: _____   Date: _____

Instructor: _____   Score: _____

Carefully read the chapter. Then answer the following questions.

PART I

1. Prepare a sketch showing the operating principle of the milling machine.

2. A column and knee type milling machine worktable has three movements. Name them.

   _____

   _____

3. Name the three basic types of column and knee type milling machines described.

   (a) Is similar to the machine in question two but a fourth table movement has been added to enable it to cut helical shapes.

   (b) Has the spindle in a vertical position.

   (c) Has a horizontal spindle and the worktable has three movements.

   3. (a) _____

   (b) _____

   (c) _____

4. List five safety precautions that must be observed when operating a milling machine.

1) _____

_____

2) _____

_____

3) _____

_____

4) _____

_____

5) _____

_____

Match the definition with the correct term.

_____ 5. The shank and body of this type cutter are made in one piece.

_____ 6. There is a suitable hole in this type cutter so it can be arbor mounted.

_____ 7. The surface being machined is parallel with the face of the cutter.

_____ 8. Has either a straight or taper shank.

_____ 9. Teeth on this type cutter are brazed or clamped in place.

_____ 10. The surface being machined is parallel with the external surface of the cutter.

_____ 11. Can be mounted directly to the spindle nose of the machine, or on a stub arbor.

(a) Face milling
(b) Peripheral milling
(c) Solid cutter
(d) Inserted tooth cutter
(e) Arbor cutter
(f) Shank cutter
(g) Facing cutter

12. Make sketches of climb milling and conventional milling.

Climb Milling   Conventional Milling

Name_____

13. Identify the milling cutters shown below. Write the name of each cutter in the space provided.

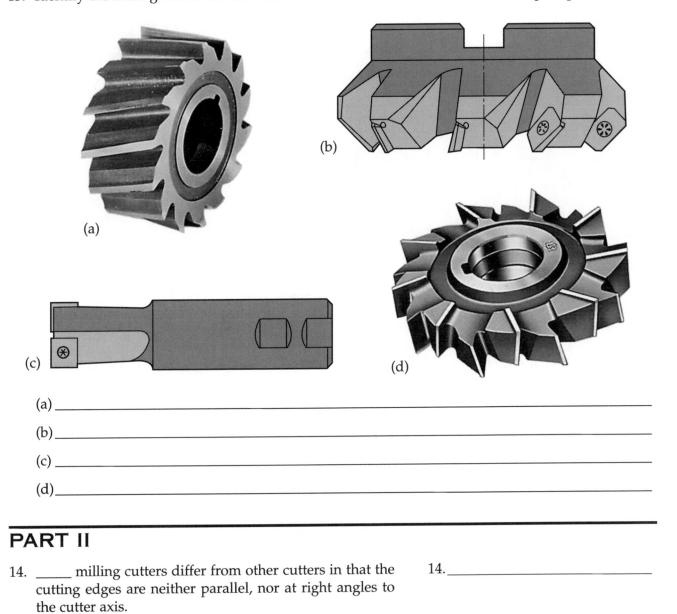

(a)

(b)

(c)

(d)

(a) _____

(b) _____

(c) _____

(d) _____

PART II

14. _____ milling cutters differ from other cutters in that the cutting edges are neither parallel, nor at right angles to the cutter axis.

14. _____

15. What are metal slitting saws? For what types of operations are they used? _____

_____

_____

_____

_____

_____

_____

16. Milling cutters are very expensive and are easily ruined if care is not taken in their use and storage. List five precautions that will extend the useful life of a milling cutter.

1) _____

_____

2) _____

_____

3) _____

_____

4) _____

_____

5) _____

_____

PART III

Using the formulas and information in Section 30.7.3 in the text, solve the following problems. Use the space provided for your work. For HSS cutters, round answers off to the nearest 50 rpm.

17. Calculate the cutting speed (rpm) of a 6″ dia. side milling cutter (HSS) to machine free cutting steel.

18. Determine the proper speed and feed for a 6″ dia. side cutter (HSS) with 16 teeth, milling brass.

19. Calculate the cutting speed (rpm) of a 1″ dia. end mill (HSS) to machine aluminum.

Name_____

20. Cutting fluids are used when milling most metal. What purposes do they serve? _____

_____

_____

_____

_____

_____

Match the definition with the correct term.

_____21. Permits rapid positioning of material for angular work.

_____22. Most widely used method of holding work for milling.

_____23. Permits compound or double angles to be machined without complex or multiple setups.

_____24. Used to divide the circumference of a round section of material into equally spaced divisions.

_____25. Vise that can be rotated or pivoted on a horizontal plane at any angle to the horizontal spindle.

_____26. Mounts directly to the worktable with jaws paralleled or at right angles to the machine's spindle.

_____27. Used to perform a variety of operations such as cutting segments of circles, circular slots, locating angularly spaced holes, and slots.

(a) Vise
(b) Flanged vise
(c) Swivel vise
(d) Toolmaker's vise
(e) Rotary table
(f) Index table
(g) Dividing head

28. _____ means using several cutters to machine multiple surfaces in one pass.

28._____

PART IV

29. Explain how to center a side cutter on round stock for the purpose of machining a keyway. Use the paper strip technique.

_____

_____

_____

_____

_____

_____

_____

30. How are hole positions centered for drilling on a vertical milling machine? _____

_____

_____

_____

_____

_____

_____

METAL SPINNING

Name: _____ **Date:** _____

Instructor: _____ **Score:** _____

Carefully read the chapter. Then answer the following questions.

PART I

1. Metal spinning is a method of _____.
    (a) weaving metal wire into cable
    (b) improving a metal's machining characteristics
    (c) working metal into three dimensional shape
    (d) All of the above.
    (e) None of the above.

2. Why is a segmented chuck used for spinning some types of complex bulged shapes? _____

    _____

    _____

    _____

3. A lubricant is needed when spinning to _____.
    (a) reduce the pressure needed to shape the metal
    (b) attach the metal disc to the chuck so spinning can be started
    (c) prevent the tailstock center from heating up
    (d) All of the above.
    (e) None of the above.

Match the definitions with the correct term.

_____ 4. Used to spin simple forms.

_____ 5. Required when spinning a bulged form (like a teapot).

_____ 6. Often made from hickory hammer handles.

_____ 7. Chuck made in several sections.

_____ 8. Fitted with pins so pressure can be applied to shape the metal disc.

_____ 9. Needed to hold metal disc on the chuck.

(a) Spinning tools
(b) Tool rest
(c) Solid chuck
(d) Follow block
(e) Segmented chuck
(f) Joined chuck

10. Prepare a sketch of a joined chuck.

11. How can pieces that have started to buckle or wrinkle during the spinning operation be straightened?

_____

_____

_____

_____

_____

_____

12. What is the difference between conventional spinning and shear spinning? _____

_____

_____

_____

_____

_____

_____

_____

Name_____

13. How is shear spinning accomplished? _____

_____

_____

_____

_____

_____

14. List five safety precautions that must be observed when metal spinning.

1) _____

_____

_____

2) _____

_____

_____

3) _____

_____

_____

4) _____

_____

_____

5) _____

_____

_____

PART II

15. Spin a simple form.

16. Make a segmented chuck for a job you have designed.

17. Spin an object using a segmented chuck.

COLD FORMING METAL SHEET

Name: _____   **Date:** _____

Instructor: _____   **Score:** _____

Carefully read the chapter. Then answer the following questions.

PART I

Match the following sentences with the appropriate word or phrase.

_____ 1. Term used for many press forming operations.

_____ 2. Process where metal sheet is cut or sheared to size for machining.

_____ 3. Used to bring sheet and plate stock to specified size or shape.

_____ 4. Equipment for cutting straight edges.

_____ 5. Process that cuts flat sheet to the shape and size of the finished part.

_____ 6. Technique in which flat metal blanks are made into three-dimensional shapes.

_____ 7. Operation used to punch openings in a material.

_____ 8. Technique used for forming shallow parts using a rubber pad to force the metal over a form.

_____ 9. Similar to the above process but is used for deep drawing.

_____ 10. A rubber diaphragm replaces the rubber mats and pads used in the two preceding processes.

_____ 11. Rubber is used to transmit the pressure needed to expand the metal blank or tube against the die.

_____ 12. Metal blank is gripped on opposite edges with clamps, then lightly pulled, forcing the metal to wrap around a form of the desired shape.

_____ 13. Shapes flat metal sheet by passing it through a series of rollers.

(a) Stretch forming
(b) Squaring shears
(c) Piercing
(d) Guerin process
(e) Hydroforming
(f) Stamping
(g) Cutting
(h) Bulging
(i) Marform process
(j) Roll forming
(k) Shearing
(l) Drawing
(m) Blanking

PART II

14. Develop a simple design for small tray, coaster, etc., and produce it by the Guerin process. Use an arbor press or vise to apply the pressure in forming the metal (use offset printing plates or heavy aluminum foil) to your design.

15. Design and manufacture equipment to demonstrate tube bending. Use hardwood forming dies.

EXTRUSION PROCESSES

 33

Name: _____ **Date:** _____

Instructor: _____ **Score:** _____

Carefully read the chapter. Then answer the following questions.

PART I

1. _____ extrusion is the process in which the piece of metal remains stationary while a hollow die stand forces the die back into the cylinder.

   (a) Impact
   (b) Direct
   (c) Indirect
   (d) Hot

   1._____

2. _____ is an extrusion process in which parts are formed by striking a slug of metal with a punch moving at high velocity.

   (a) Impact
   (b) Direct
   (c) Indirect
   (d) Hot

   2._____

3. In _____ extrusion, the ram and product move in the same direction against the die.

   (a) impact
   (b) direct
   (c) indirect
   (d) hot

   3._____

4. What limits the size of an extrusion? _____

   _____

   _____

5. In cold extrusion operations, _____.

   (a) the punch is cooled to near freezing temperatures
   (b) the ingot is usually left at room temperature
   (c) the billet is cooled to near freezing temperatures
   (d) All of the above.
   (e) None of the above.

   5._____

147

6. Prepare a sketch showing how to make extrusions that are wider than the diameter of the die.

PART II

7. Secure samples of metal parts made by the extrusion process.

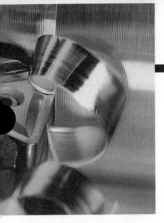

POWDER METALLURGY

Name: _____ Date: _____

Instructor: _____ Score: _____

Carefully read the chapter. Then answer the following questions.

PART I

1. What is *powder metallurgy?* _____
   _____
   _____
   _____
   _____

2. There are many applications for powder metal parts. List five uses.

   1) _____

   2) _____

   3) _____

   4) _____

   5) _____

3. What is the first phase in the manufacture of powder metal products? _____
   _____
   _____
   _____

4. To change a briquette into a strong, useful unit it must be   4._____
   _____.

5. Which operation involves reworking sintered parts in a die similar to the original compacting die?
   _____
   _____
   _____

PART II

6. Secure samples of parts made by the powder metallurgy process. These samples could include bearings and fuel filters. Examine the powder metal parts and those parts made from solid metal (machined, cast, forged) under a microscope. How do they differ?

_____

_____

_____

_____

_____

_____

NONTRADITIONAL MACHINING TECHNIQUES

35

Name: _____   **Date:** _____

Instructor: _____   **Score:** _____

Carefully read the chapter. Then answer the following questions.

PART I

1. Electrical discharge machining (EDM) is a process in which _____.
   - (a) metals that are difficult to machine by traditional chip making techniques can be worked to close tolerances
   - (b) holes of almost any shape can be produced
   - (c) an electric spark is used to erode the metal
   - (d) All of the above.
   - (e) None of the above.

   1._____

2. Only _____ materials can be machined by the process described in the preceding statement.

   2._____

3. _____ is similar to band machining but uses a small diameter wire electrode in place of the saw.

   3._____

4. Technique is electroplating in reverse.

   4._____

5. _____ is a process that uses chemicals to etch away the metal.

   5._____

6. How does chemical milling differ from chemical blanking? _____

   _____

   _____

   _____

   _____

   _____

   _____

7. The immersion time in chemical milling must be carefully
   controlled or _____.

   (a) the metal will be vaporized
   (b) the metal surface will be too difficult to clean
   (c) too much metal may be removed
   (d) None of the above.
   (e) All of the above.

7._____

8. Involves the total removal of metal from certain areas of
   the part by chemical action.

8._____

9. List the steps in the chem-blanking process.

   1) _____

   _____

   2) _____

   _____

   3) _____

   _____

   4) _____

   _____

   5) _____

   _____

10. What is *hydrodynamic machining?* _____

   _____

   _____

11. In _____ machining, a focused electron beam vaporizes
    the material at the cutting point.

11._____

12. _____ machining uses an intense beam of light to cut the
    material.

12._____

13. In _____ machining, sound waves are used to pound a
    slurry of fine abrasive particles against the work.

13._____

14. _____ machining uses ultrasonics and a special tool to
    force abrasives against the work. No heat is generated in
    this process.

14._____

15. What is *springback?* _____

   _____

16. In which type of forming operation is metal shaped in
    microseconds using pressure generated by the sudden
    application of large amounts of energy.

16._____

Name_____

17. _____ uses a high-energy pressure pulse of an explosive   17._____
    charge to form the metal.

18. What is *lead time?* _____

    _____

    _____

19. In _____ forming, a magnetic field causes the work to   19._____
    collapse, compress, shrink, or expand depending on the
    design and placement of the coil.

PART II

Safety Note: *Under no conditions should explosive forming be attempted using fireworks or other chemical explosives.*

20. Demonstrate the chemical milling process using magnesium metal sheet and vinegar. *Wear safety glasses when doing this experiment.*

QUALITY CONTROL

Name: _____ **Date:** _____

Instructor: _____ **Score:** _____

Carefully read the chapter. Then answer the following questions.

1. What is the primary purpose of quality control?_____

   _____

   _____

   _____

   _____

2. _____ testing involves the complete damaging of the part,   2._____
   while _____ testing allows the part to be used for its   _____
   intended purpose.

3. Give an example of destructive testing._____

   _____

   _____

   _____

   _____

4. Give an example of nondestructive testing. _____

   _____

   _____

   _____

   _____

5. What is an *optical comparator?* How is it used to inspect metal parts? _____

   _____

   _____

   _____

   _____

   _____

6. Briefly describe *radiographic inspection.* _____

_____

_____

_____

_____

_____

7. What is *magnetic particle inspection?* _____

_____

_____

_____

_____

_____

_____

8. How is *fluorescent penetrant inspection* used to inspect parts?_____

_____

_____

_____

_____

_____

9. How is *spotcheck inspection* different from fluorescent penetrant inspection? _____

_____

_____

_____

_____

_____

10. Briefly describe *ultrasonic inspection.* _____

_____

_____

_____

_____

_____

NUMERICAL CONTROL AND AUTOMATION

Name: _____   Date: _____

Instructor: _____   Score: _____

Carefully read the chapter. Then answer the following questions.

1. Give a brief description of numerical control machine tool operation. _____

   _____

   _____

   _____

   _____

2. _____ instructions are used to identify each NC machine   2. _____
   function.

3. What is the *Cartesian Coordinate System?* _____

   _____

   _____

   _____

   _____

   _____

   _____

   _____

   _____

4. With incremental tool positioning, each tool movement is made from _____

   _____

   _____

   _____.

5. Absolute tool movement measures all tool movement from_____

   _____

   _____

   _____.

6. Complete the drawings to show an example of:

(a) Incremental dimensioning

(b) Absolute dimensioning

(a)

(b)

7. The point-to-point NC tool positioning system is normally used for _____.

    7._____

    (a) controlled tool movement along one axis at a time
    (b) precisely controlling tool movement at all times
    (c) normally used for simple machining operations (drilling, spot welding, punching)
    (d) All of the above.

8. Manual programming can be done _____.

    8._____

    (a) if the machining operations have a low degree of complexity
    (b) by anyone who can interpret engineering drawings
    (c) by someone with a knowledge of machine tool operations
    (d) All of the above.
    (e) None of the above.

9. When is computer-aided programming required? _____

_____

_____

_____

10. What is unique about CAD/CAM (computer-aided design/computer-aided manufacturing)?

_____

_____

_____

_____

_____

Name_____

11. Automation is an industrial technique whereby _____.   11._____

    (a) robots do all of the work
    (b) only work that is too hazardous for humans is done
        by robots
    (c) mechanical labor and help are substituted for
        human labor and control
    (d) All of the above.
    (e) None of the above.

12. _____ is often referred to as computer integrated manu-   12._____
facturing because of its flexibility.

13. Smart tooling involves cutting tools and part holding devices that can be_____

_____

_____

_____.

14. JIT (just-in-time) is a system that _____

_____

_____

_____.

15. Many types of robots have been devised. All of them consist of four basic components. What are
these components?

_____

_____

_____

Accident Report

Name: _____   **Date:** _____

Instructor: _____   **Score:** _____

1. Injured person: _____

   Address: _____

   Telephone: _____

   Homeroom: _____

2. Accident Witnesses

   Name: _____

   Address: _____

   Telephone: _____

   Name: _____

   Address: _____

   Telephone: _____

3. Type of injury: _____

   _____

   _____

   _____

4. Treatment

   First aid: _____   By whom: _____

   Physician: _____   Address: _____

   Hospital: _____   Address: _____

5. Cause of accident: _____

   _____

   _____

6. Tools/machines/equipment involved: _____

   _____

   _____

7. Action taken to prevent recurrence of accident: _____

   _____

   _____

Project Plan Sheet

Name: _____ Period:_____

Name of project: _____

Date begun: _____ Date completed: _____

Source of idea or project: _____

Bill of Material					
Part Name	No. of Pieces	Material	Size (T × W × L)	Unit Cost	Total Cost
				Total Cost	

Plan of Procedure

Name: _____   **Period:** _____

Name of Project: _____

List the operations to be performed in their sequential order. Indicate the tool(s) and equipment needed to accomplish the job.

Number	Operation	Tools and Equipment

List the difficulties encountered while constructing this project and explain how they were solved.

_____

_____

_____

_____

Student's evaluation: _____

Instructor's evaluation: _____

Instructor's comments: _____

_____

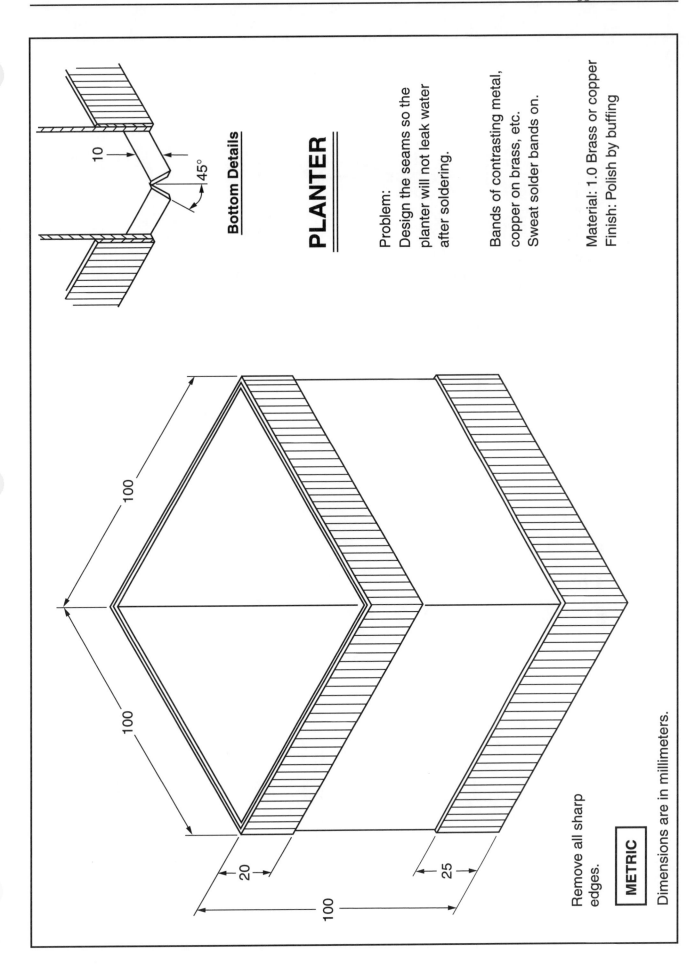

Bottom Details

PLANTER

Problem:
Design the seams so the planter will not leak water after soldering.

Bands of contrasting metal, copper on brass, etc. Sweat solder bands on.

Material: 1.0 Brass or copper
Finish: Polish by buffing

Remove all sharp edges.

METRIC

Dimensions are in millimeters.

DESIGN PROBLEM

Design and fabricate an official NBA size basketball hoop.

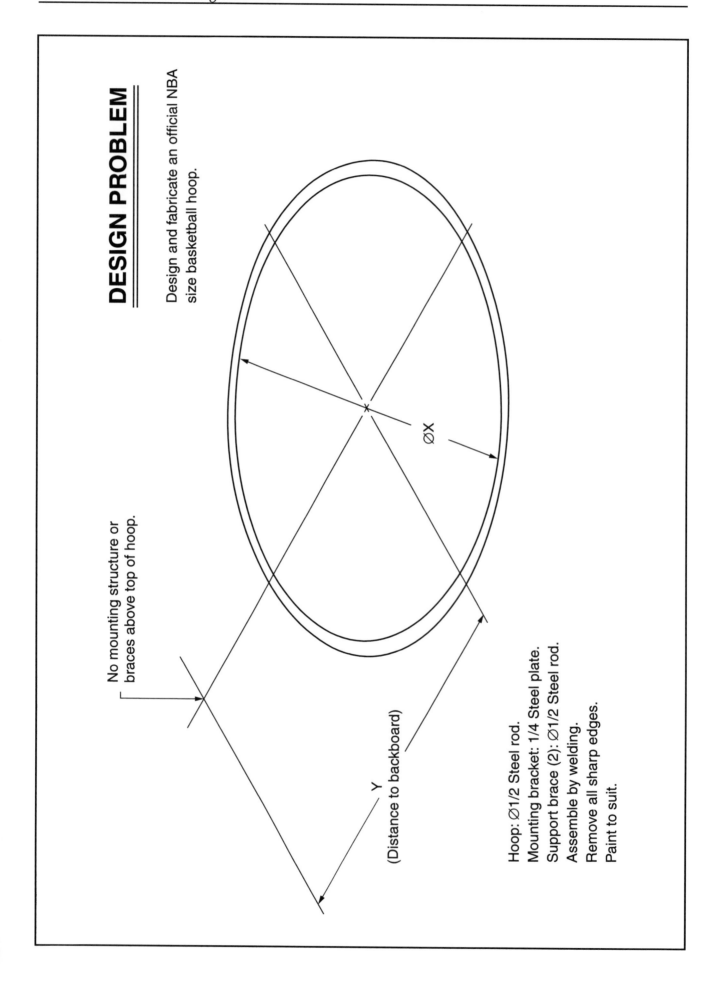

⌀X

No mounting structure or braces above top of hoop.

Y
(Distance to backboard)

Hoop: ⌀1/2 Steel rod.
Mounting bracket: 1/4 Steel plate.
Support brace (2): ⌀1/2 Steel rod.
Assemble by welding.
Remove all sharp edges.
Paint to suit.

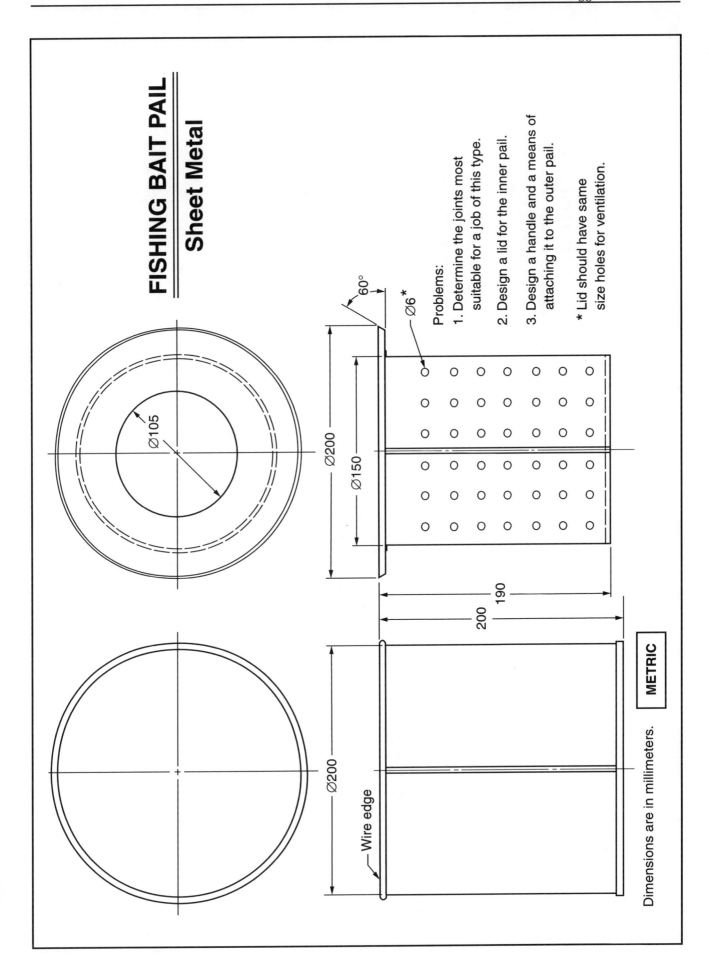

FISHING BAIT PAIL
Sheet Metal

60°

Ø6 *

Problems:

1. Determine the joints most suitable for a job of this type.

2. Design a lid for the inner pail.

3. Design a handle and a means of attaching it to the outer pail.

* Lid should have same size holes for ventilation.

Ø105

Ø200

Ø150

190

200

Ø200

Wire edge

Dimensions are in millimeters.   **METRIC**

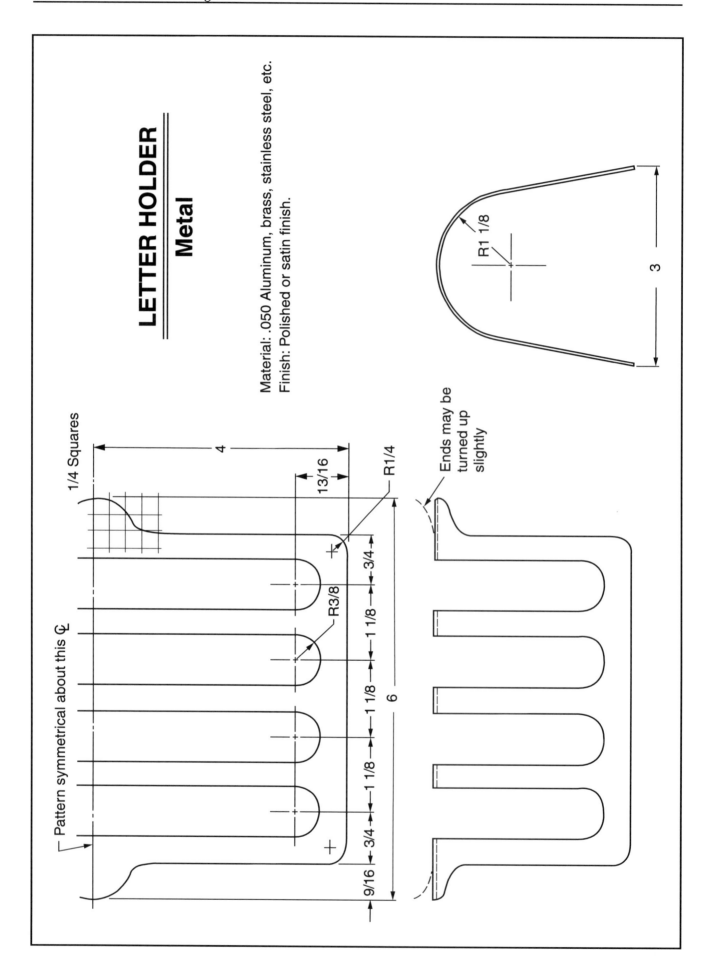

LETTER HOLDER
Metal

Material: .050 Aluminum, brass, stainless steel, etc.
Finish: Polished or satin finish.

R1 1/8

3

1/4 Squares

Pattern symmetrical about this ℄

4

13/16

R1/4

R3/8

3/4

1 1/8

1 1/8

1 1/8

3/4

9/16

6

Ends may be turned up slighty

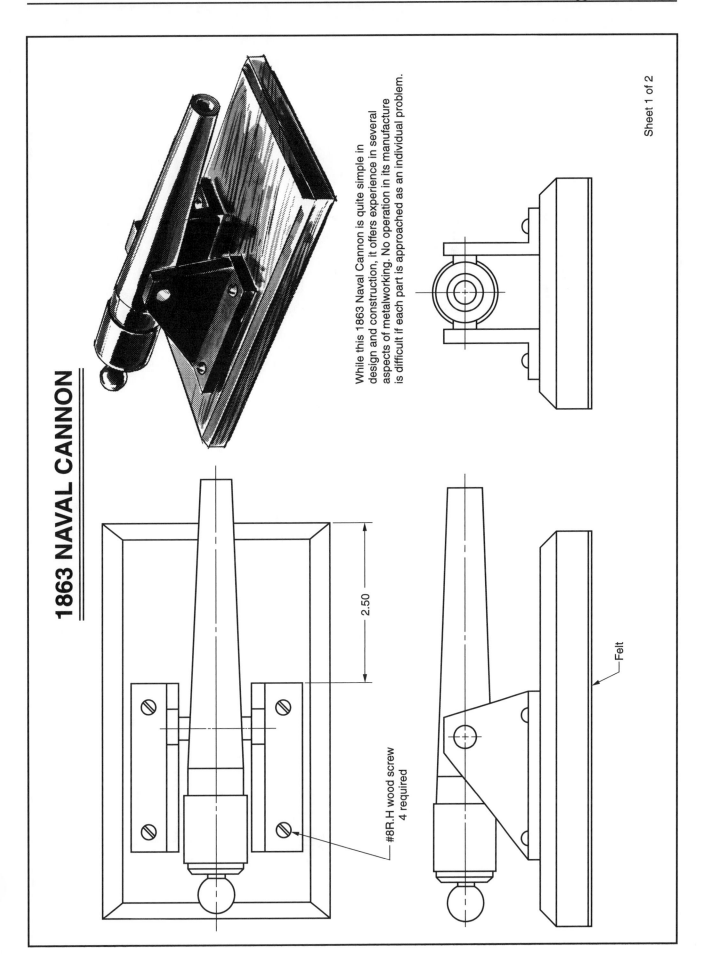

1863 NAVAL CANNON

While this 1863 Naval Cannon is quite simple in design and construction, it offers experience in several aspects of metalworking. No operation in its manufacture is difficult if each part is approached as an individual problem.

Sheet 1 of 2

2.50

#8 R.H wood screw
4 required

Felt

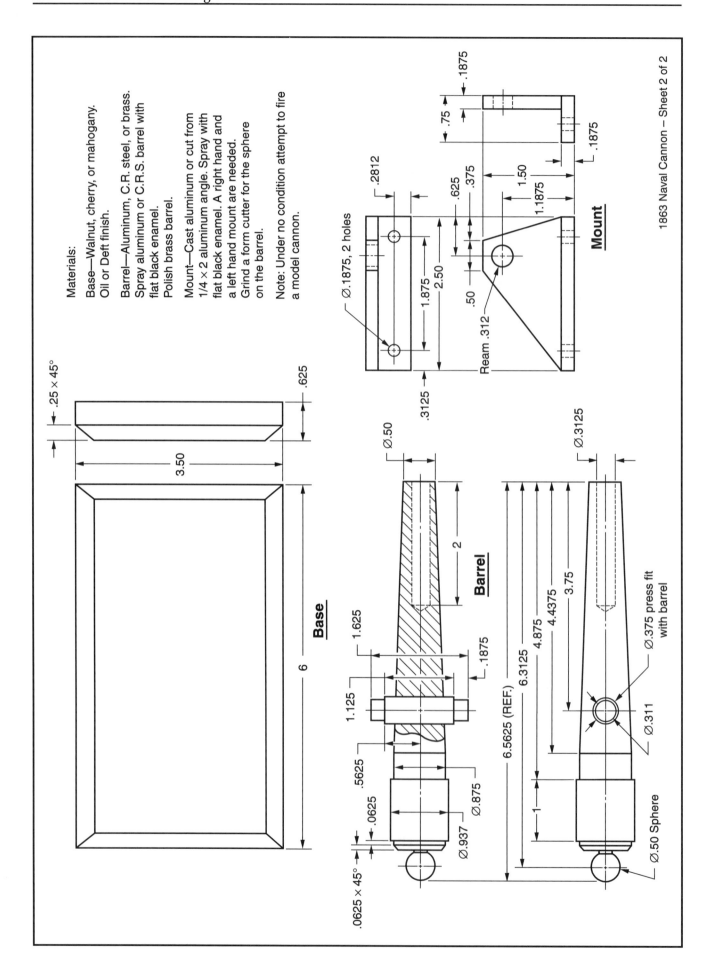

Materials:

Base—Walnut, cherry, or mahogany. Oil or Deft finish.

Barrel—Aluminum, C.R. steel, or brass. Spray aluminum or C.R.S. barrel with flat black enamel. Polish brass barrel.

Mount—Cast aluminum or cut from 1/4 × 2 aluminum angle. Spray with flat black enamel. A right hand and a left hand mount are needed. Grind a form cutter for the sphere on the barrel.

Note: Under no condition attempt to fire a model cannon.

Base

Barrel

Mount

1863 Naval Cannon – Sheet 2 of 2

SHELF BRACKETS

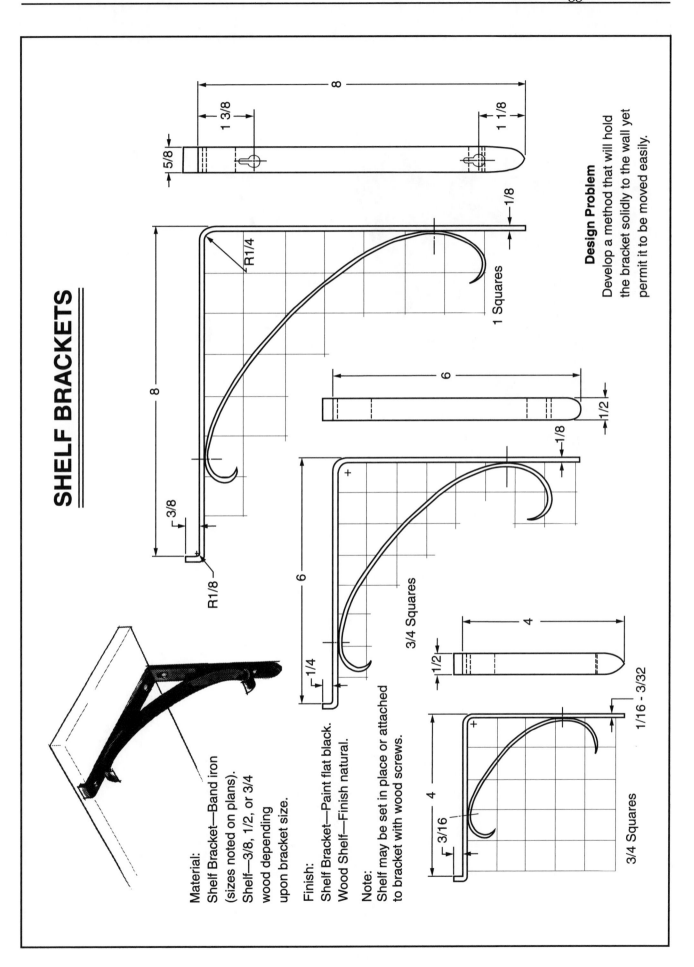

Design Problem

Develop a method that will hold the bracket solidly to the wall yet permit it to be moved easily.

8

1 3/8

1 1/8

5/8

R1/4

8

1 Squares

1/8

R1/8

3/8

6

1/2

1/8

6

1/4

3/4 Squares

4

1/2

1/16 - 3/32

4

3/16

3/4 Squares

Material:
Shelf Bracket—Band iron (sizes noted on plans).
Shelf—3/8, 1/2, or 3/4 wood depending upon bracket size.

Finish:
Shelf Bracket—Paint flat black.
Wood Shelf—Finish natural.

Note:
Shelf may be set in place or attached to bracket with wood screws.

SHOE SCRAPER

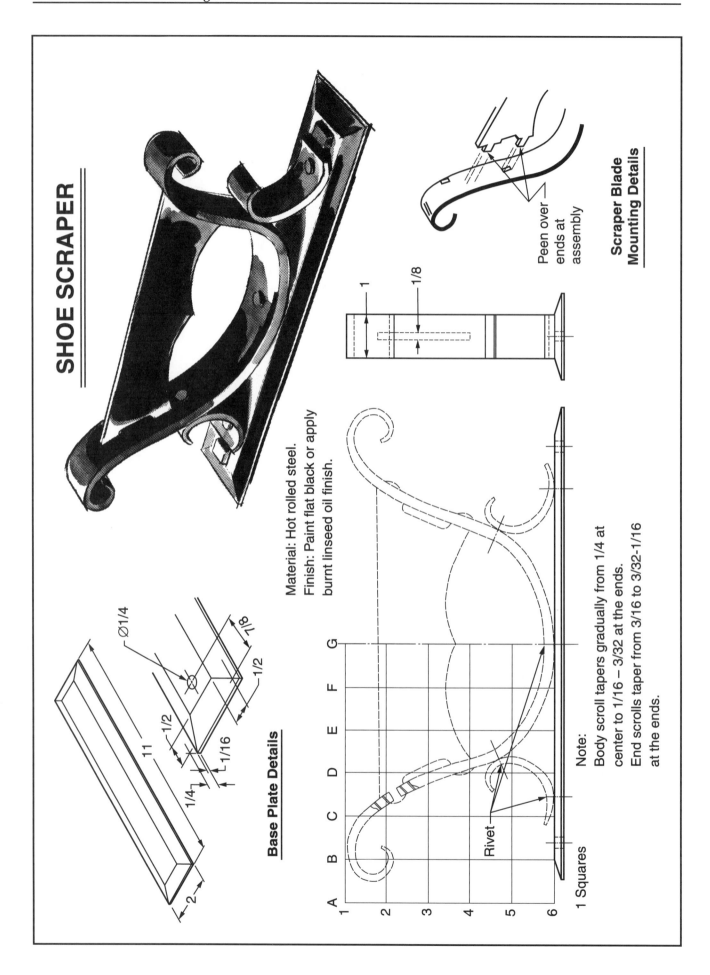

Material: Hot rolled steel.
Finish: Paint flat black or apply burnt linseed oil finish.

Base Plate Details

Ø1/4

11

2

7/8

1/2

1/4

1/2

1/16

1

1/8

Scraper Blade Mounting Details

Peen over ends at assembly

Rivet

Note:
Body scroll tapers gradually from 1/4 at center to 1/16 – 3/32 at the ends.
End scrolls taper from 3/16 to 3/32-1/16 at the ends.

A   B   C   D   E   F   G

1
2
3
4
5
6

1 Squares

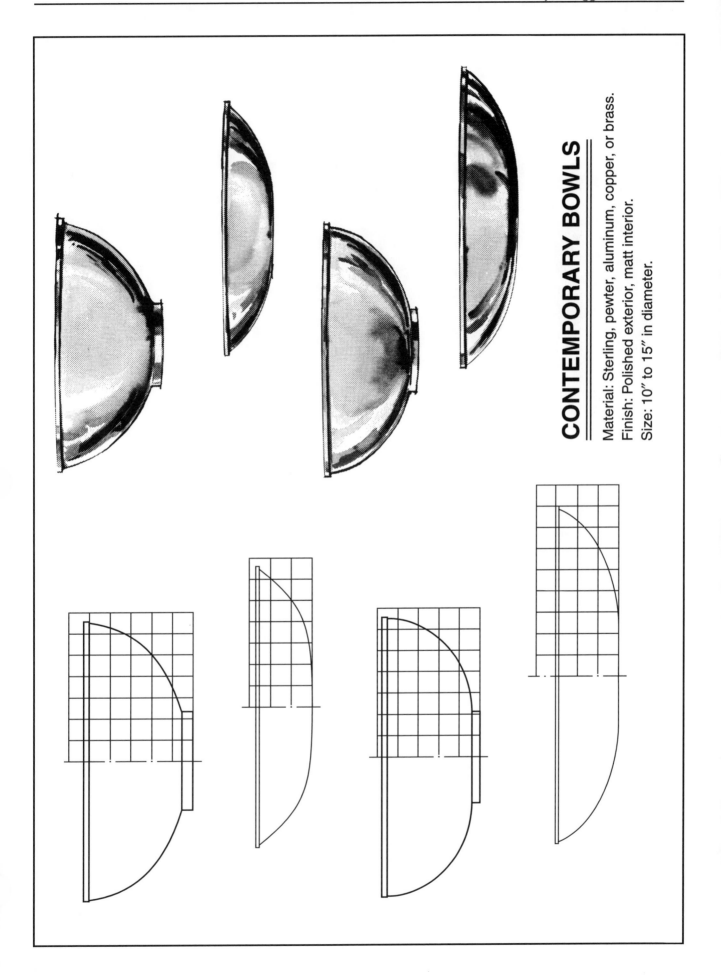

CONTEMPORARY BOWLS

Material: Sterling, pewter, aluminum, copper, or brass.
Finish: Polished exterior, matt interior.
Size: 10" to 15" in diameter.